Dairy Cattle
Feeds and Feeding
in Western Canada

Steve Mason

©AgroMedia International Inc.
'Science into practice'

ISBN 978-1-7772967-2-8

Acknowledgements:

The author acknowledges the financial support of the Westgen Endowment Fund for the production and distribution of this book.

Attributable credits:

Tables 6.3, 6.4 and all tables in appendices A and B are adapted from NASEM Nutrient Requirements of Dairy Cattle, 2021 with permission. Figure 1.1: licensed Alamy stock photo; Figures 1.4, 1.5, 1.8 and 1.9: Norman Criddle in G.H. Clark and M.O. Malte, Fodder and Pasture Plants, Canada Department of Agriculture, 1923; Figure 6.12: adapted from Edmonson. A. J. et al., J. Dairy Sci. 72:68-78 (1989). Where credits for other tables and images do not appear in their captions, they are either original, in the public domain or otherwise unattributable.

About the author:

Steve Mason obtained both his B.Sc. (Biochemistry) and Ph.D. (Animal Science) from the University of British Columbia. Following a post-doctoral fellowship in medical pharmacology at the University of Calgary, he became a livestock producer for several years which led to stints as Provincial Livestock Specialist and later, Provincial Livestock Nutritionist, with the BC Ministry of Agriculture.

Moving back to Alberta, Steve became Manager of ProLivestock Nutrition/Management Specialists, a consulting unit of United Grain Growers, later Agricore United, where he provided dairy nutrition expertise to feed mill sales staff and their clients. Subsequently, after a short assignment as Senior Extension Associate with Cornell University's Pro-Dairy program, Steve established AgroMedia International Inc., a business that provides knowledge translation and transfer services as well as contract scientific and technical support to the Canadian dairy industry. Operating under the name 'AgInformatics' the company also provides data management and analysis services to the livestock research community. More recently, a partnership operating as 'Farm Animal Care Associates' has focused on assisting dairy producers with the adoption of best management practices for animal health and welfare in the context of Dairy Farmers of Canada's proAction initiative. Since 2010, Steve has served as an Adjunct Associate Professor with the University of Calgary Faculty of Veterinary Medicine, teaching nutrition, mentoring graduate students and participating in research. He is a Registered Professional Animal Scientist (PAS) and a Diplomate of the American College of Animal Nutrition.

Disclaimer:

Neither the author nor AgroMedia International Inc. make representation or warranty of any kind, express or implied, regarding the accuracy, adequacy, validity, reliability, availability or completeness of the information provided.

To register your copy of this publication for notification of updates, to obtain a copy of the Excel-based dairy ration formulator described in chapter 7 or to order additional copies in either print or digital (pdf) format, contact:
AgroMedia International Inc.
2508 Charlebois Drive NW
Calgary AB T2L 0T6 Canada
(403)807-5404
steve@agromedia.ca

Table of Contents

Introduction

Dairy farming in Canada has evolved significantly over the past few decades. As our country's population has become more diverse and dietary preferences have shifted, the demands for particular dairy products have also changed. For example, while margarine was previously thought to be nutritionally superior to butter, that belief it is now understood to be incorrect with the result that butter consumption has experienced a resurgence. This has led to the requirement for dairy farmers to increase the butterfat content of the milk they produce by adjusting herd genetics and feeding management.

Consumers have also influenced other aspects of herd management through their increased concern about animal welfare. This has led to significant revisions to guidelines for the care and handling of animals, including increasing attention to providing adequate nutrition, especially to young calves.

While the evolution of the industry in response to consumer demands has shaped a number of changes, automation, the adoption of new technologies and ongoing research into dairy cattle genetics and nutrition have also influenced management practices.

The current 'state-of-the art' in dairy nutrition is contained in the recent revision of the *Nutrient Requirements of Dairy Cattle*, the latest in a series first published in 1944. Up to and including the 2001 7th edition, the series had been published by the US National Research Council (NRC) and were commonly referred to as 'NRC Dairy'. In 2015, the NRC was renamed the National Academies of Sciences, Engineering, and Medicine (NASEM) so the 8th edition, published in December 2021, is referred to as NASEM Dairy 8.

Each edition has been what is now called a 'Consensus Study Report'—the result of a collaboration of a number of leading researchers in the field of dairy nutrition who have agreed upon a set of nutrient requirement recommendations based on their review of all of the currently available data published in peer-reviewed scientific journals. Although Canadian researchers have contributed significantly to the current body of knowledge about dairy cattle nutrition, the greater proportion has originated in the US where feeds and feeding practices are shaped by an industry that is somewhat different from that in western Canada.

This guide attempts to place dairy cattle feeds and feeding practices in the context of the western Canadian dairy industry, where:

- production is constrained by the Canadian dairy supply management system, requiring producers to limit their production to that permitted by their quota holdings;

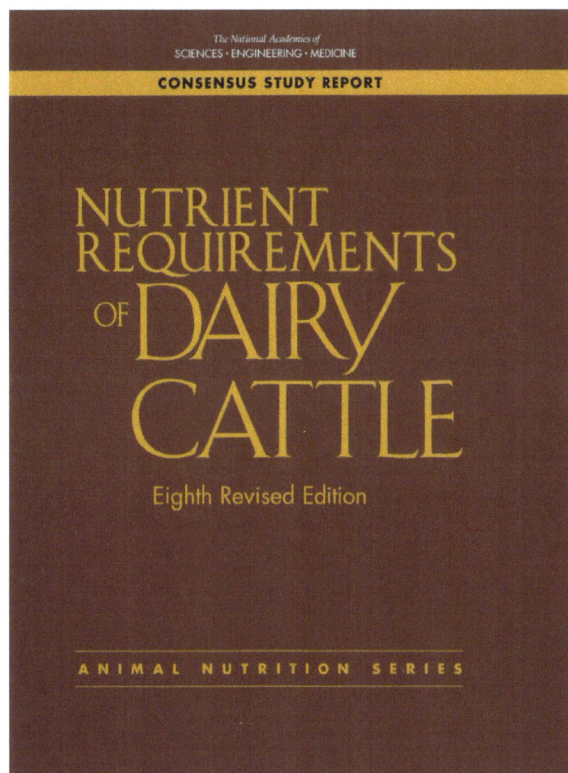

- herd size is limited by the cost of quota—at the time of writing (July 2022), the weighted average quota price across the 4 western provinces was $39,799/kg, compared with $24,000/kg in the eastern provinces;
- in spite of the higher cost of quota, the average size of herds enrolled in Dairy Herd Improvement (DHI) in 2021 was 177 lactating cows in western Canada versus 84 across provinces east of Manitoba;
- mandatory upper limits on the ratio of solids non-fat to butterfat (SNF:BF) in shipped milk require producers to adjust feeding strategies in favour of increasing their butterfat tests;
- few western Canadian dairy herds provide prolonged pasture access to their lactating cows due to high land values in south coastal BC and to short growing seasons on the prairies;
- the principle feed grain fed to dairy cattle in western Canada is barley—while corn is most commonly fed in the US, few locations in western Canada receive enough heat during the growing season to ripen corn;
- although corn silage is commonly fed in south coastal BC and in some areas on the southern prairies (as it is in most of the US), the predominant grain crop silage fed to western Canadian herds is barley silage;
- of herds enrolled in DHI in 2021, western Canada had a significantly lower proportion of herds managed in tie stalls and fed individually than those in the eastern provinces (8.6% vs 68.6%) and a higher proportion of herds with automated milking systems (25.4% vs 13.6%).

Chapter 1: Feeds

Cattle are herbivores—plant-eaters—so a logical starting point for a discussion on cattle feeding is a description of the plants and plant constituents that they consume.

Plant growth and composition

The growth of plants

Plant growth is summarized in figure 1.1. All green plants absorb carbon dioxide (CO2) from the air through their leaves. Water (H_2O), nitrates and minerals are assimilated from the soil by the roots. Sunlight, which is trapped by the green pigment chlorophyll, provides the energy that is used by the plant to make carbohydrates, proteins and other organic constituents from these simple precursors.

Carbohydrates

Carbohydrates are combinations of carbon dioxide and water. The simplest combinations, mono- and disaccharides, (commonly called simple sugars) include glucose, galactose, fructose and ribose. These water-soluble carbohydrates are readily transported throughout the plant to provide for a variety of metabolic requirements.

Mono- and disaccharides are the building blocks for more complex carbohydrate polymers (polysaccharides) such as cellulose, hemicellulose, pectins, pentosans and starch.

Plant carbohydrates are often classified as structural or nonstructural. As the name implies, structural carbohydrates (SC) provide structural stability for upright growth; they make up the more fibrous parts of the plant. Carbohydrates of this type are found in the cell wall and include: cellulose, hemicellulose and

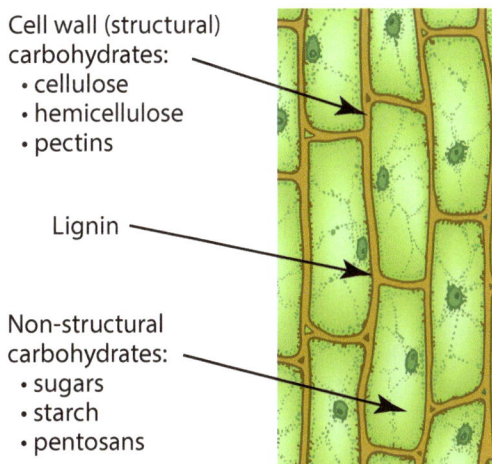

Leaves absorb carbon dioxide (CO_2) from the air. In the presence of sunlight, they combine CO_2 and water (H_2O) to form carbohydrates.

Roots absorb water, nitrates and minerals from the soil.

H_2O
Nitrates
Ca, P, Mg, etc.

Figure 1.1: Assimilation of substrates in plant growth.

pectin. Non-structural carbohydrates (NSC; also called non-fibre carbohydrates, NFC) are primarily found in plant cell contents and include the simple sugars and storage carbohydrates such as pentosans and starch.

Lignin

Lignin is the cell wall component that gives plants the rigidity they require to stand upright. As plants mature and increase in height, the proportion of lignin in their cell walls increases, particularly in stems and other structural parts. In addition to being totally indigestible, lignin inhibits microbial degradation of cell wall carbohydrates, as depicted in figure 1.3.

Fats and oils

Like carbohydrates, fats and oils are composed of carbon, hydrogen and oxygen. However, the proportion of carbon and hydrogen is much greater with the result that fats and oils furnish about 2.25 times as much energy per kilogram than carbohydrates and proteins.

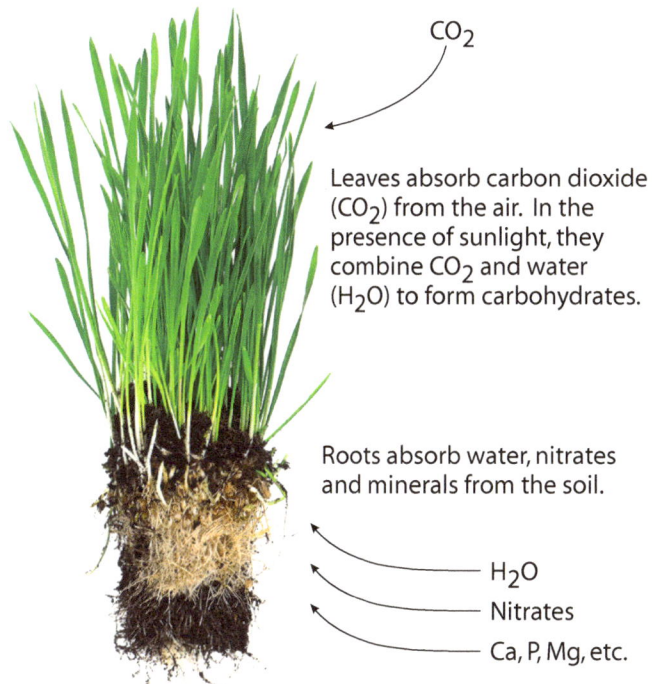

Cell wall (structural) carbohydrates:
• cellulose
• hemicellulose
• pectins

Lignin

Non-structural carbohydrates:
• sugars
• starch
• pentosans

Figure 1.2: Structural carbohydrates and lignin are found in the cell wall; non-sructural carbohydrates are found in seeds and in the cytoplasm of the growing plant.

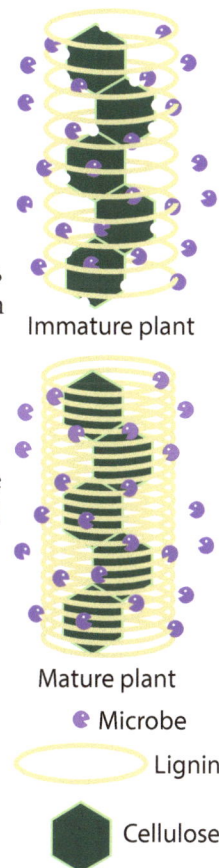

Immature plant

Mature plant

🦠 Microbe

◯ Lignin

⬡ Cellulose

Figure 1.3: As plants mature, increasing lignin content inhibits microbial degradation of structural carbohydrates.

	Lignin, % of dry matter
Alfalfa	
fresh, early bloom	7.0
fresh, midbloom	9.0
fresh, full bloom	10.0
hay, sun-cured, late bloom	12.0
hay, sun-cured, mature	14.0
leaves only, sun-cured	5.0
Corn	
grain	0.9
cobs	5.9
silage, average	2.6
Barley	
grain	2.0
silage, headed	5.6
straw	11.0

Table 1.1: Lignin concentrations vary with plant species, stage of maturity and plant component.

Plants, except for the oilseeds, contain relatively low concentrations of fats and oils (collectively called lipids). Unlike animals, which store energy as fat, plants store energy as carbohydrate, as described above.

Proteins

In addition to carbon (from CO_2), hydrogen (from H_2O) and oxygen (from both), plant proteins contain nitrogen and most contain sulfur and phosphorus assimilated from the soil. Mainly functioning as enzymes in plants, their number and variety are almost infinite. Proteins are predominantly found in the reproductive and actively growing parts such as leaves. In animals, protein is found mainly in the form of muscle tissue. And, since milk protein is one of the primary products of the dairy enterprise, it is easy to appreciate the importance of protein in the feed.

Legume forages such as alfalfa and clover generally contain higher levels of protein than grasses. This is due to the large supply of nitrogen available to them through fixation from the atmosphere. Nitrogen fixation is facilitated by bacteria (rhizobia) contained in nodules on the roots which transfer nitrates to the plant (figure 1.4). In exchange, the legume plant supplies soluble carbohydrates which provide energy to the bacteria.

	Ether Extract, % of dry matter
Alfalfa	
fresh, early bloom	3.1
hay, sun-cured, late bloom	1.8
Canola	
meal, solvent extracted	1.8
whole seeds	40.5
Corn	
grain	4.2
silage, average	3.2
distillers grains with solubles	10.0
Barley	
grain	2.1
silage, headed	3.5
Grasses, cool season	
hay, immature	3.3
hay, mature	2.0
silage, mid-maturity	2.4
Soy	
meal, solvent extracted	1.1
whole beans	19.2

Table 1.2: Lipid concentrations vary with plant species, stage of maturity and plant part. Lipid content is estimated by extraction with ether.

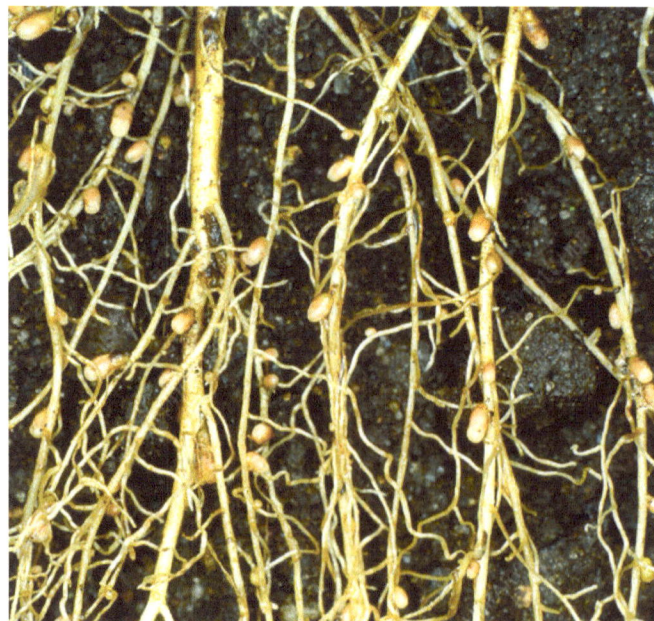

Figure 1.4: Rhizobia root nodules on white clover.

	Crude Protein, % of dry matter
Alfalfa	
fresh, early bloom	22.0
fresh, midbloom	18.0
fresh, full bloom	14.0
Canola	
meal, solvent extracted	40.0
Corn	
grain	9.4
distillers grains w solubles	29.7
silage, average	8.8
Barley	
grain	12.5
silage, headed	12.0
Grasses, cool season	
hay, immature	18.0
hay, mature	10.8
silage, mid-maturity	16.8
Soy	
meal, solvent extracted	53.8

Table 1.3: Crude protein concentrations in various feeds measured as nitrogen concentration divided by 16%—the average nitrogen content of common feed proteins.

Vitamins

Like their other organic constituents, plants synthesize vitamins from raw materials absorbed through their roots and leaves. Some of them, like the B-vitamins serve much the same functions in plants as they do in animals. Others serve a specific function in plants which is quite different from their function in animals. For example, β-carotene is a yellow plant pigment that plays a part in photosynthesis. Animals convert β-carotene to vitamin A which is essential in maintenance of epithelial (surface) tissues.

	vitamin A equivalents IU/kg DM
Fescue pasture	39,865
Fescue hay	2,923
Alfalfa hay	2,913
Orchardgrass hay	3,109
Wheat straw	60
Corn silage	6,900
Corn grain	170
Soybean meal	55
Corn gluten meal	3,747

Table 1.5: Typical content of vitamin A equivalents in various feeds. source: C.L. Pickworth et al., J Anim Sci 90:1553 (2012)

Plants synthesize all of the vitamins and vitamin precursors (e.g., β-carotene is the precursor of vitamin A) required in the diets of higher animals. However, in most situations, ruminant diets need only contain the fat-soluble vitamins A, D and E, due to the fact that rumen microbes synthesize the full range of water-soluble vitamins (see page 62). However, the concentrations of A, D and E in plant-based feeds can be extremely variable, typically declining rapidly after harvest due to oxidation.

Minerals

Plants assimilate minerals from the soil through their roots. Although present in relatively small amounts, they are as essential to the development of the plant as they are to the animals that consume them. Minerals are well distributed in the plant, largely occurring in association with the organic compounds (carbohydrates, proteins, lipids and vitamins). For example, magnesium (Mg) is an essential component of chlorophyll; cobalt (Co) is an essential component of vitamin B12; potassium (K) is a co-factor for many plant enzymes.

	Macro-minerals, % of DM				Micro-minerals, mg/kg DM			
	Na	Mg	Ca	P	Cu	Mn	Zn	Se
Legume hay	0.17	0.30	1.82	0.24	11.0	31.0	24.0	0.54
Cool season grass hay	0.04	0.20	0.58	0.23	9.0	72.0	31.0	0.06
Corn grain	0.02	0.12	0.04	0.30	3.0	11.0	27.0	0.07
Barley grain	0.02	0.14	0.06	0.39	6.0	22.0	38.0	0.11
Canola meal	0.11	0.60	0.70	1.20	6.4	57.8	64.4	1.20
Soymeal	0.03	0.29	0.35	0.70	16.0	40.0	58.0	0.13

Table 1.4: Example macro- and micro-mineral concentrations in common feeds. DM: dry matter.

Feeds

Forages

Forage feeds consist primarily of the vegetative parts (i.e., stems and leaves) of plants that contain relatively higher concentrations of structural carbohydrates compared to the reproductive parts (e.g., seeds, tubers) that contain higher levels of non-structural carbohydrates and other nutrients.

The nutritional value of forages is largely determined by their stage of maturity as illustrated in figure 1.5. As plants grow and mature, stems and leaves accumulate lignin which is not only indigestible itself, but also forms complexes with structural carbohydrates, inhibiting their digestion. At the same time, the protein content of the plant declines.

Perennial grass forages

Perennial grasses commonly utilized in western Canadian dairy cattle diets as hay or haylage include:

Bromegrass: there are two types of bromegrass—a southern type and a northern type. In western Canada, the northern type (smooth bromegrass) is more vigorous and has higher seed yields than the southern type (meadow bromegrass). It is tolerant of drought and extreme temperatures and can be grown alone or mixed with other grasses and legumes. Bromegrass is the most commonly used companion to alfalfa in mixtures grown on dryland (i.e., without irrigation).

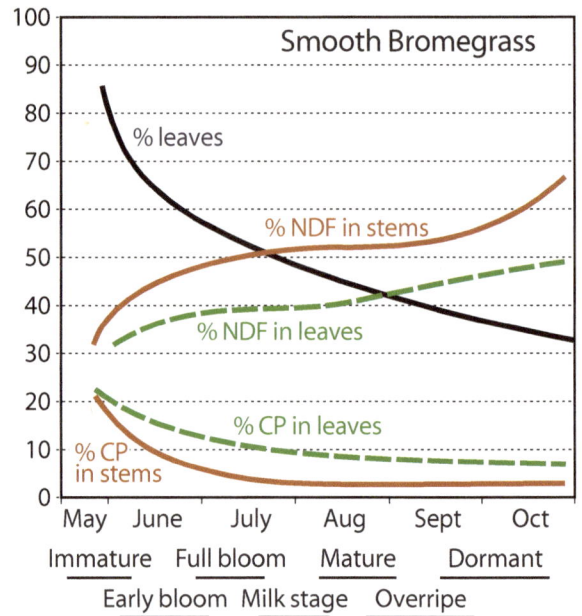

Figure 1.5: Composition changes as a grass crop matures. CP: crude protein. NDF: neutral detergent fibre.

Orchardgrass is used for pasture, silage and hay. It has a deep, competitive root system and can interfere with nutrient uptake in legumes so the two are not normally grown together, especially when heavily fertilized with nitrogen.

Perennial Ryegrass is rarely grown in the prairie provinces although it is common in south coastal British Columbia. A short-lived bunch grass with a shallow root system, it is very palatable and nutritious when harvested at the correct maturity.

Figure 1.6a: Perennial grasses at maturity: left - smooth bromegrass; centre - orchardgrass; right - perennial ryegrass.

Tall Fescue can grow even on the poorest of soils. It is tolerant of both acidic and alkaline soils and, although it has a low moisture requirement, tall fescue does especially well under moist conditions like those found in south coastal British Columbia where it is often seeded as a mix with orchardgrass and/or ryegrass.

Reed Canarygrass is particularly tolerant of low, poorly drained areas. Canarygrass is palatable to cattle as long as it is not allowed to become too mature as it becomes coarse with age. It is often used to stabilize the banks of waterways because of its ability to develop a dense, erosion-resistant sod.

Figure 1.6b: Perennial grasses at maturity: left - tall fescue; right - reed canarygrass.

Timothy is used for both pasture and hay; it grows well with legumes; and is easy to harvest. Overmature timothy is very fibrous, limiting intake.

Wheatgrass: there are several different types of wheatgrass: slender, crested, intermediate, pubescent, streambank and tall. These plants are persistent, drought-resistant, and can be found throughout the Canadian prairies.

Annual grass forages

Other than the grain crop forages described on the following pages, annual grasses are not commonly used in western Canadian dairy diets. The one exception is *Annual (Italian) Ryegrass*, a fast growing species that establishes easily and provides high yields when soil moisture is adequate. Although most commonly used for grazing, Italian Ryegrass makes good quality hay or silage when harvested at the correct stage of maturity. Changes in composition with advancing maturity are similar to those for perennial grasses.

Grass identification

Be aware that, for ease of identification, these images of grass forages show them at maturity, well beyond the stage at which they would be harvested for feed. As illustrated in figure 1.5, at this stage of growth the nutritional value of the vegetative parts of the plant has reached its lowest level. However, it is often difficult to accurately differentiate grass species in their earlier vegetative stages without resorting to close inspection of their finer anatomic features.

Figure 1.6d: Annual (Italian) Ryegrass at maturity.

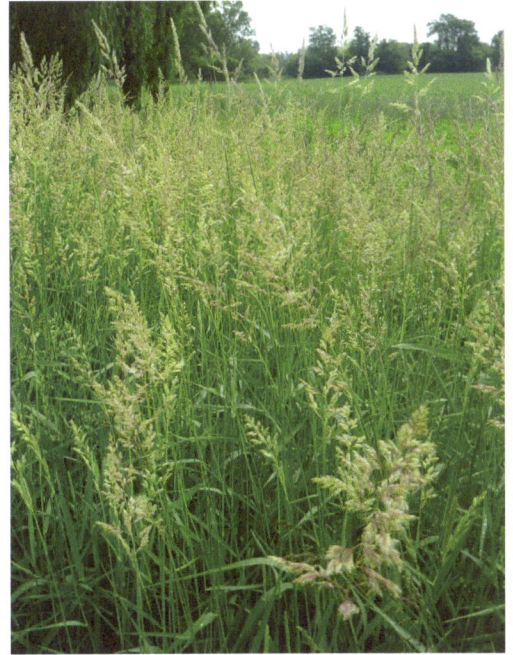

Figure 1.6c: Perennial grasses at maturity: left - timothy; right - wheatgrass.

Grain (cereal) crop forages

The 'small' grain crops (members of the grass family, *Gramineae*), including barley, rye, oats, triticale and wheat, are common forage sources in western Canadian dairy diets. These are harvested for silage when the grain is in the early development (late milk to soft dough) stage or for dry feed at the 'boot' stage, when the immature seed head emerges from the leaf sheath. When utilized after field drying, these forages may be referred to as greenfeed; when harvested and stored wet, they are often referred to as grain-crop silage.

Barley grows best in well-drained, fertile soils. This plant is also adapted to growth in sandy soils. The disadvantage to using barley is the presence of spiky awns on the seed heads which decrease palatability.

Rye is well suited for pasture because of its high productivity. A disadvantage of using rye as pasture is that it quickly becomes unpalatable as it matures.

Oats can be grazed or stored as hay (greenfeed) or silage. Particular varieties (e.g., Foothills) have been bred for use as high quality forages.

Triticale is the result of a cross between wheat and rye. It has the potential to be higher yielding than barley but many attempts to use triticale silage in lactating dairy cow diets have been disappointing due to low intakes.

Wheat can yield quality pasture, greenfeed or silage that is palatable, high in protein and energy potential and low in fibre.

When ensiling these small grains, the moisture content should be 62–68% to limit excessive air in the stems and higher concentrations of butyric acid in the silage. As pasture, small grain crops can be grazed during winter months (where weather permits) and into early spring without reducing later harvest yields.

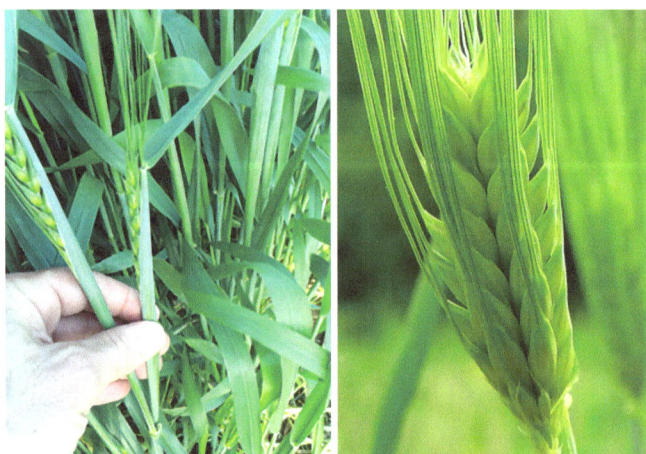

Figure 1.7: Small grain crops should be cut for hay at the boot stage (left); for silage they should be harvested when the grain is in the late milk to soft dough stage.

Corn is commonly grown for silage in south coastal British Columbia but often will not achieve optimum maturity when grown on the prairies due to shorter, cooler growing seasons. When harvested for silage, the crop should be cut when it reaches a dry matter level of 30–35% (65–70% moisture). Corn silage is highly palatable and a good source of digestible energy (due to its grain content) but low in digestible protein.

Perennial legume forages

Legume forages are primarily fed as hay and silage. They are seldom offered to ruminants as pasture or as fresh-cut forage because of their potential to cause bloat. When ensiled, perennial grasses and legumes are often called hay-crop silages or haylage to differentiate them from grain-crop (cereal) silages.

Like grasses, the intake potential and nutritional quality of legumes are largely dependent on their stage of maturity. As the plants mature, their stems elongate, the proportion of leaf to stem decreases, fibre content increases and protein declines. These relationships are illustrated in figure 1.8. In comparison to grasses, legumes are higher in protein, calcium, magnesium, sulfur and copper, but lower in manganese and zinc.

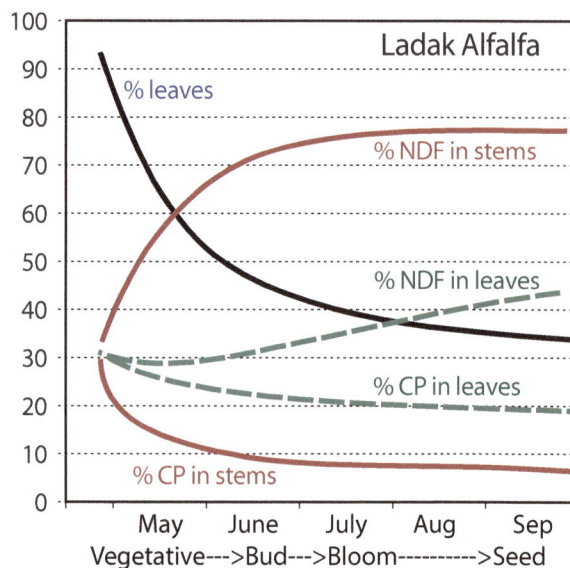

Figure 1.8: Composition changes as alfalfa crop matures. CP: crude protein. NDF: neutral detergent fibre.

Alfalfa is a perennial legume that can survive in both cold and warm climates, typically providing 2 to 4 cuts of hay or silage per year, depending on soil moisture and the length of the growing season. It has excellent drought tolerance due to its deep roots and established stands have fair salt tolerance but the crop cannot tolerate acidic soil or a high water table. Grazing or feeding recently harvested alfalfa can cause bloat in ruminants.

Figure 1.9a: Perennial legumes in bloom: left - alfalfa; centre - sainfoin; right - birdsfoot trefoil.

Sainfoin, native to Europe and parts of Russia and Asia, was introduced into western Canada in the 1960s as a non-bloating alternative to alfalfa. It is used to a limited extent for hay and pasture, although it has poor longevity and poor tolerance to grazing.

Birdsfoot Trefoil is a perennial legume adapted to temperate to cold climates. It is tolerant of drought and acidic soils and can be used for hay or pasture. Trefoil is palatable and nutritious but also has the potential to cause bloat.

Cicer Milkvetch is a perennial that can be grown in cold climates. It is tolerant of saline soil, drought, flooding and acidic and alkaline conditions. Because the crop does not cause bloat, it is commonly used for pasture; less frequently for hay.

Red Clover is a perennial that can be grown in moist conditions in cold, warm and acidic soils. It will, however, not tolerate drought. There are two types of red clover: single-cut and double-cut. Single cut varieties flower later, are better able to withstand winter and are larger. Red clover of both types are most suitable for silage; used as pasture there is a risk of bloat.

Figure 1.9b: Red clover in bloom.

Concentrates

Concentrate ingredients generally contain lower concentrations of fibre relative to those classified as forages. As a result, most concentrates yield higher levels of nutrients per unit fed when digested by both animals and intestinal microbes. Concentrates include grains, oilseeds, protein sources, fats and oils, mineral and vitamin supplements, and additives.

With the exception of grains, most small dairy herd operators may not be directly exposed to the other concentrates described below except when they are included in commercial mixed feeds and supplements. Producers managing larger herds may purchase individual concentrate ingredients as bulk commodities to be blended into complete diets on-farm.

Grains

Grains are the seeds of the cereal crops described above for use as forages, including barley, rye, oats, triticale, wheat and corn. To be efficiently digested by cattle, grains must be processed by either grinding or rolling (see pages 26 - 27). This is because the fibrous hull of the whole grain kernel prevents access to its starchy core by both rumen microbes and mammalian digestive enzymes. Conversely, over-processing exposes the starchy endosperm to rapid degradation by rumen microbes—an advantage in terms of maximizing microbial protein synthesis but a disadvantage due to a higher risk of ruminal acidosis resulting from a high rate of volatile fatty acid production.

Figure 1.10: Small grains commonly used in western Canadian livestock diets. B - barley; O - oats; R - rye; W - wheat.

Barley is the principle feed grain used in western Canadian dairy diets. After processing, barley starch is one of the more rapidly degraded among the common grains, as indicated in table 1.6.

Oats grain contains less starch and more fibre than the other feed grains and, therefore, contributes less energy to the diet than an equal quantity of other grains. At one time, this was considered an advantage—oats grain was fed in preference to other 'hotter' grains because it posed a lower risk of acidosis. With its fibrous hull, oats require processing before feeding but rolling or grinding produces a relatively bulky ingredient.

Rye is similar to barley in its appearance and use as a dietary energy source. It is palatable, but that palatability may decrease with inclusion rates exceeding about 20% of the grain mix. Rye is susceptible to ergot, a fungus which can produce a number of mycotoxins (see page 19). Grain containing ergot is indicated by the presence of large and misshapen kernels (figure 1.11) and should not be fed.

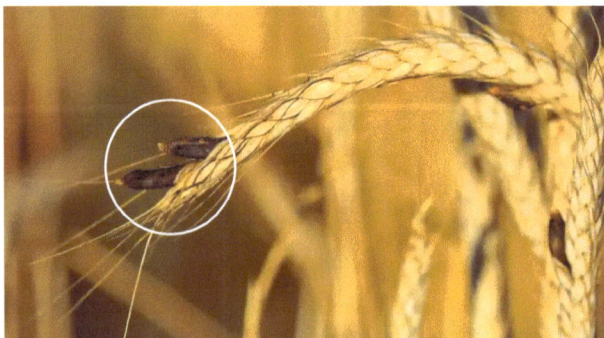

Figure 1.11: Ergot on a rye seed head.

Wheat usually has a higher protein content than either barley or corn. Although its energy value is roughly equivalent to that of corn, processed wheat is very rapidly degraded in the rumen (see table 1.6),

Grain Component	Wheat	Barley	Corn
	Degradation Rate, %/hr		
Dry Matter	28.6	17.8	11.5
Starch	34.1	30.9	8.6

Table 1.6: Rumen degradation rates of processed grains.

increasing the risk of acidosis when fed in large quantities (e.g., more than 25% of diet DM). When wheat grown for human consumption fails to meet grade standards, it is often graded as feed wheat which becomes available for livestock feeding at prices competitive with barley and other feed grains.

Wheat midds are a by-product of the wheat milling process, composed of varying proportions of wheat bran, shorts, germ, flour, and other residual parts of the grain. As a result of this variability, chemical composition can also be widely variable. Wheat midds are commonly used in commercial mixed feeds as a source of energy, fibre and phosphorus when their price is competitive with alternative sources.

Corn: Little corn is harvested as grain in western Canada due to inadequate heat during short growing seasons. However, processed corn grain is often included in diets for cows in early lactation, taking advantage of its palatability and its slower rate of rumen degradation (table 1.6). The lower rate of volatile fatty acid production lowers the risk of acidosis which may reduce feed intake.

The degradation rates in table 1.6 apply to dry rolled or ground grains. Steam rolling, steam flaking and extrusion increase the starch degradation rates of all grains. This effect is most pronounced for corn as moist heat opens its crystalline starch structure to rumen microbial degradation.

Figure 1.12: Canola seed (left) and meal (right).

Oilseeds and meals

Canola is widely grown in western Canada for its high quality oil. If effectively processed, whole canola seeds can be incorporated into lactation diets. But, unless the right equipment is used, rolling is difficult because of the high (40%+) fat content of the seeds. Finely grooved rollers equipped with scrapers are necessary to keep the oily, crushed product from accumulating on the rollers. The high levels of unsaturated fatty acids in canola oil can reduce fibre digestion by rumen microbes, limiting its inclusion rate to about 2% of diet DM. The total dietary inclusion rate of added fats and oils should not exceed 5 - 6% of DM.

Canola meal, a by-product of oil extraction from whole seeds, is widely used as a protein source in livestock diets. In lactation diets, canola meal can be included at up to 10% of diet dry matter, typically as an ingredient in a commercial concentrate mix.

Cottonseed: Although two types of cottonseed are available in Canada—whole (fuzzy) and delinted—most of the seed used in dairy diets is whole, due to the potential contribution of its residual linters (figure 1.13) to effective fibre requirements.

Figure 1.13: Whole (fuzzy) cottonseed. The residual 'fuzz' (linters) are pure cellulose.

Cottonseed hulls are commonly used in the US as a source of dietary fibre in dairy lactation diets. Cottonseed is also an excellent source of rumen undegradable protein and energy due to its relatively high oil content (~20% of DM). When fed whole, the seed coat at least partially inhibits the highly unsaturated oil from causing negative effects on rumen fermentation. When stored on-farm, cottonseed should be protected from moisture and be well ventilated to reduce the risk of moulds that may produce gossypol, a potent mycotoxin (see page 19).

Flaxseed (linseed) contains approximately 30 - 40% linseed oil of which about 50% is alpha-linolenic acid, an omega-3 fatty acid which is believed to reduce the risk of heart disease in humans. In several research trials at the University of Alberta, feeding flaxseed to lactating cows has increased the milk concentrations of fatty acids considered beneficial for human health while decreasing the concentrations of several others considered detrimental. Rolled flaxseed had a greater effect than whole seed.

Figure 1.14: Flaxseed, also known as linseed.

Because linseed oil is high in unsaturated fatty acids, no more than 1.5 kg of rolled flaxseed or 2.5 kg of whole flaxseed should be fed per day in lactation diets. Larger amounts can interfere with rumen fermentation and reduce feed intake.

Soybeans are a heat- and moisture-loving crop. Although extensively grown in Ontario and Québec, only in the most southerly parts of the prairie provinces are growing seasons warm enough for routine cultivation. However, if the price is right, raw or roasted soybeans can be successfully fed to lactating cows at up to 10% of dietary dry matter.

Soymeal is the product remaining after extracting oil from whole soybeans. It is high in protein and energy and is one of the most commonly used protein supplements in livestock diets. Its excellent

Figure 1.15: Soybeans (left) and soymeal (right).

palatability makes it particularly suitable as a protein supplement in calf starter diets.

Sunflower seeds: Two varieties of sunflowers are grown in parts of Canada which have sufficiently long growing seasons. A lower-oil variety is grown for birdseed and direct human consumption. A higher-oil variety is grown for sunflower oil production and may be fed to lactating cows as a supplemental energy source. The high-oil seeds contain 35 - 45% oil and do not require rolling before feeding.

Like canola and linseed oils, sunflower oil contains high concentrations of unsaturated fatty acids which can alter rumen fermentation and reduce feed intake. Therefore, a maximum daily intake of 1.8 kg of whole seeds is recommended in lactation diets.

Figure 1:16: Sunflower seeds.

Other protein sources

Peas: Field pea production in western Canada has increased dramatically over the past 30 years, making them readily available for inclusion in dairy diets. Five different types of peas that are commonly available in western Canada are shown in figure 1.17. Large and small, green and yellow peas are the most common.

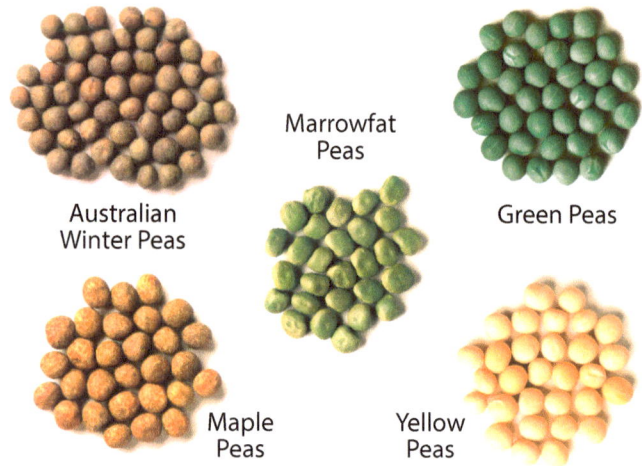

Figure 1.17: Field pea varieties commonly available in western Canada.

The nutrient composition of peas can be widely variable. Samples of 27 pea varieties tested by Alberta Agriculture varied in crude protein content from 23 to 27% of dry matter. In samples of 5 varieties tested at the University of Saskatchewan feed lab, starch values ranged from 27 to 50%. Therefore, when including peas in diets, it is important to base inclusion rates on analysis of each load. The ruminal degradation rate of pea starch is similar to that of corn starch (table 1.6) while the rapid degradability of pea protein may limit the inclusion rate of peas in lactation diets.

Brewer's grains, commonly referred to as brew mash, are a byproduct of the beer brewing industry. Because the cost of drying is high relative to its feeding value, brewer's grains are most commonly available as a wet product containing 20 - 25% dry matter (75 - 80% moisture). The high moisture content makes long distance transportation uneconomic and prolonged storage problematic. Therefore, farms that are able to use brewer's grains are generally located within about 150 km of a brewery and must be prepared to take delivery of the product at least every 2 weeks.

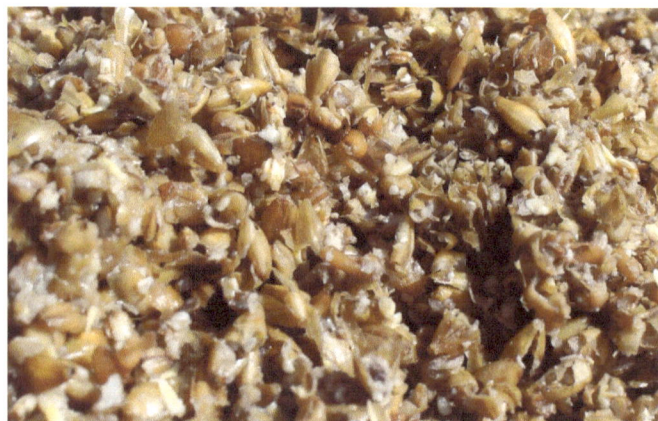

Figure 1.18: Wet brewer's grains.

In storage, the moisture content of brew mash decreases steadily as effluent runs off and significant heating and deterioration can occur, particularly in warm weather. Therefore, it should be stored in a commodity bin with concrete sides and floor, a drain for effluent collection and a close-fitting plastic cover. The product can be successfully ensiled in small batches if provision is made to minimize effluent loss and prevent surface exposure to air.

Brewer's grains are a good source of digestible fibre and rumen undegradable protein (RUP). Lactating cows should not be fed more than about 3.5 kg of dry matter from this by-product, the main limitation being its moisture content. High moisture (greater than about 55%) rations can limit total dry matter intake.

Distiller's grains are the by-product from the yeast fermentation of grain to produce ethanol. At the conclusion of the fermentation process, the liquid phase containing the ethanol is separated from the solid grain residue. The dried residue is dried (or dehydrated) distiller's grains (DDG). In most distilleries, after distilling off the ethanol, the remaining liquid is combined with the solid residue and dried. The resulting product is called dried (or dehydrated) distiller's grains with solubles (DDG/S).

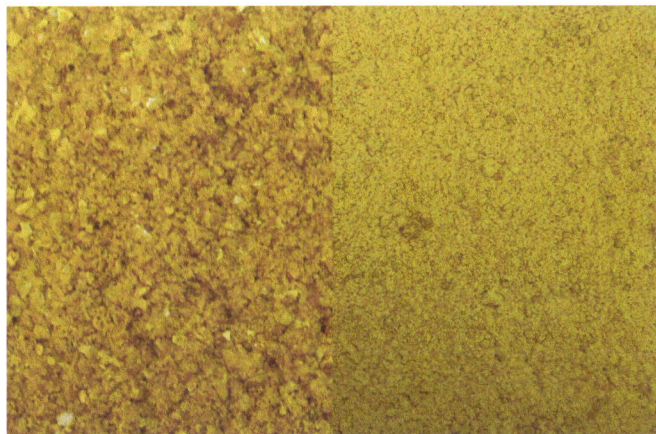

Figure 1.19: Corn distiller's grains (left) and corn gluten meal (right).

Most of the DDG and DDG/S in western Canada is the product of corn distillation for the production of whiskey. This is a high quality product of uniform golden brown colour, containing about 8% fat and 25% crude protein with relatively high bypass (RUP) value. Its palatability and the high digestibility of its fibre (40 – 45% NDF) make it a valuable ingredient in diets for lactating cows.

Corn DDG/S imported from the US is usually a by-product of the fuel ethanol industry. Although this product often appears to contain burnt particles that might indicated heat-damaged protein (see page 34),

there is no evidence that it is of inferior feeding quality. Wheat and rye DDG and DDG/S are also sometimes available in western Canada.

If DDG comprise a large proportion of the diet, their high phosphorus levels demand calcium supplementation to maintain an adequate dietary Ca:P ratio (see page 59).

Corn gluten meal is a by-product of the manufacture of corn starch and corn syrup. It is often used as a source of bypass protein (RUP) in lactation diets but has earned a reputation of being of lower quality due to its low lysine content and inferior digestibility.

Corn gluten meal should not be confused with *corn gluten feed* which is an entirely different by-product, not readily available in western Canada.

Other plant by-products

Beet pulp is the residue remaining after the extraction of sugar from sugar beets. When molasses is added back to the pulp after processing, the product is called molassed beet pulp or beet pulp with molasses. The addition of molasses increases the product's protein concentration slightly, reduces its fibre content, raises its sugar content and improves its palatability. Beet pulp is commonly available as 'shreds' or pellets.

Traditionally, dry or moistened beet pulp has been used as a 'top-dress' on forage or other mixed feeds to stimulate intake. Its high fibre content, the high digestibility of that fibre and its palatability make beet pulp a valuable ingredient in automated milking system 'pellets' and complete lactation diets. This is particularly true when lower quality forages increase the dietary requirement for grain. At these times, beet pulp can be used to replace a portion of the grain, reducing the risk of acidosis and often improving total feed intake. Wet beet pulp has also been used to replace corn silage in diets, serving as a source of both roughage (physically effective fibre) and energy.

Figure 1.20: Beet pulp shreds (left) and pellets (right).

Animal by-products

Blood meal is a by-product of the animal slaughter and rendering industries. For the production of blood meal, blood is either batch-dried, spray-dried or ring-dried. Batch drying reduces the quality and digestibility of the protein and, although this product will often be less expensive, it should not be expected to yield the response produced by spray- or ring-dried meal. Blood meal is valuable in animal diets as a concentrated source of high quality protein—rich in essential amino acids (other than isoleucine) and high in rumen undegradable protein for lactating dairy cows. Although it is quite palatable for dairy cattle, blood meal should not be fed to lactating cows at a level higher than 0.3 kg/day.

Meat and bone meal is the dried and rendered product from mammalian tissues not including horn, hair, hide trimmings, manure, stomach contents, added blood meal or poultry by-product. Although commonly included in dairy lactation diets prior to the discovery of bovine spongiform encephalopathy (BSE; mad cow disease), meat and bone meal from cattle is no longer allowed in ruminant diets (see information about specified risk materials above) although the porcine product is still permitted.

Feather meal is produced using high pressure steam to hydrolyze feathers removed from poultry at slaughter. Without hydrolysis, digestibility is very low. Feather meal is sometimes used in dairy lactation diets as an inexpensive source of relatively poor quality rumen undegradable protein. The poorer protein quality is due to a low level of lysine, although the meal is rich in the essential sulfur-containing amino acids, methionine and cysteine. Maximum recommended feeding rate is 3% of dietary dry matter.

Fish meal: a by-product of the fish packing industry, good quality fish meal is an ideal source of protein for non-ruminants and rumen undegradable protein (RUP) for ruminants. In a review of 12 years of RUP feeding trials, only fish meal produced consistently positive increases in milk production. These results were likely due to its high content of the essential amino acids, lysine and methionine.

Although it is usually assumed that the palatability of fish meal will reduce dietary intake, this is seldom a problem when high quality meal has been kept dry to prevent oxidative rancidity. Nevertheless, it is a good idea to introduce fish meal into the diet gradually, if possible. Good quality fish meal may be included in diets for all livestock species and companion animals, although its cost may be a limitation. Fish meal is usually quite expensive in western Canada due to demand by the fish farming industry.

In lactation diets, fish meal is usually fed at no more than about 0.7 kg/cow/day. Because the oil content of various fish meals can vary widely, a common recommendation is to feed no more meal than will provide 100 grams per day of fish oil.

Poultry by-product meal consists of the ground, rendered, clean parts of the carcasses of slaughtered poultry, such as heads, necks, feet, undeveloped eggs, and intestines, exclusive of feathers except in such amounts as might occur unavoidably in good processing practices.

Poultry meal consists of the clean combination of rendered poultry flesh and skin with or without bone. It does not contain feathers, heads, feet or entrails. If the meal is from a particular source, it may state so (e.g., chicken, turkey, etc). Depending on the raw products used, storage and processing, crude protein content can be quite variable. Because of its significant content of unsaturated fatty acids, oxidative rancidity can be an issue if antioxidants are not included in the meal.

Due to its low palatability, the inclusion rate of poultry meal in dairy cattle diets is usually restricted to less than 1 kg/day as part of a total mixed diet.

Fats and oils

Several by-product fats and oils are commonly added to early lactation diets to increase energy intake at a time when cows are in negative energy balance. Products which are liquid at room temperature (20°C) are oils; those that are solid are fats. Melting point is an indication of the degree of saturation of the fatty acids contained in the product. Vegetable oils contain more unsaturated fatty acids; animal fats contain more saturated fatty acids.

Unsaturated fatty acids are converted to saturated fatty acids in the rumen (they are *biohydrogenated*) and may reduce feed intake by interfering with fibre digestion. Therefore, it is recommended that no more than 0.5 kg/d of raw oil (i.e., not contained in oilseeds) be included in diets for lactating cows.

As illustrated in table 1.7, the long chain fatty acid (see table 1.7 and page 53) composition of fats and oils (lipids) that may be included in dairy diets is widely variable. Commonly used lipids include:

Oilseed oils (canola, flaxseed, sunflower, safflower) may be used in feeds as energy sources and to keep dust levels down in very dry concentrate mixtures.

Tallow is a byproduct of the animal slaughter and rendering industries. Because it contains more saturated fatty acids, it can be included at levels up to 2% of dietary dry matter in dairy lactation diets (roughly 0.5 kg/cow/day), subject to SRM restrictions on impurities (less than 0.15%; see page 14).

Yellow grease, also called feed fat, is a mixture of refuse fats and oils from the restaurant industry. An appropriate level of inclusion in dairy lactation diets is no more than 1.5% of dietary dry matter.

Bypass fat products, including calcium salts of long-chain fatty acids and highly saturated long-chain fatty acids in prilled or flaked form may be included in dairy lactation diets at levels of up to 0.7 kg/cow/day.

Lipid	Fatty acid composition, %			
	16:0	18:0	18:1	18:2
Canola oil	4.8	1.6	53.8	22.1
Flaxseed oil	5.3	4.1	20.2	12.7
Soybean oil	10.3	3.8	22.8	51
Sunflower oil	5.4	3.5	45.3	39.8
Tallow	24.5	19.3	40.9	3.2
Ca salt PFAD	51.0	4.0	36.0	7.0
C16:0 enriched	89.7	1.0	5.9	1.3

Table 1.7: Nutritionally important long chain fatty acids in common lipid supplements. XX:Y – XX = fatty acid carbon chain length, Y = number of double bonds. PFAD: palm fatty acid distillate.

The limitations for use of these products are price and the declining digestibility of long-chain fatty acids in the cow's small intestine as total lipid intake increases (see page 53).

It has become common practice to feed 0.3 - 0.5 kg/d of palm oil or its derivatives, containing high proportions of palmitic acid (16:0). Although this strategy is very effective in increasing milk fat test, since 2020 it has attracted criticism due to concern about the environmental degradation resulting from the expanding world palm oil industry.

The maximum total lipid content of a dairy lactation diet should not exceed 6.5 - 7% of dietary dry matter (DM), recognizing that the base ingredients in the diet will provide 3 - 4% even without the addition of supplementary fats or oils. Typically, rumen available lipids (oilseed oils, tallow, yellow grease) and rumen inert (bypass) products will each provide an additional 1.5 - 2% of diet DM.

Commercial mixed feeds

The feed ingredients described above are some of the most commonly fed to dairy cattle in western Canada. However, examination of the ingredient list for any of the commercial mixed feeds available in Canada may reveal the presence of several others. Although a feed manufacturer will provide a guaranteed analysis for each product it sells, the ingredient content of each of those products will vary from time to time, depending primarily on current prices and availability.

All animal feeds that are imported or manufactured and sold in Canada are subject to the *Feeds Act* and *Feed Regulations*, administered by the Canadian Food Inspection Agency (CFIA). Schedules IV and V of the *Feeds Regulations* list all feed ingredients approved for use in Canada including detailed descriptions and requirements for their use (see figure 1.21).

SCHEDULE IV PART I	
CLASS 4. ENERGY FEEDS	
4.3 Seed and Mill Screenings	
4.3.7	Pulse grain screenings refuse (or Refuse screenings pulse grains) (IFN --) means pulse grain screenings (i.e. from chickpeas, lentils, peas and beans, solely or a mixture thereof) conforming to the Refuse Screenings standard referred to in the Off Grades of Grain and Grades of Screenings Order. If it bears a name descriptive of kind (e.g. lentil grain screenings) or form (i.e. pelleted) the product shall correspond thereto. If any pelleting aid(s) is used, the common name or names shall be indicated on the label.
CLASS 5. PROTEIN FEEDS	
5.1 Animal	
5.1.7	Animal meat and bone meal rendered (or Meat and bone meal) (IFN 5-00-388) is a product obtained by rendering animal tissues, exclusive of hair, hoof, horn, hide trimmings, manure and stomach contents except in such amounts as may occur unavoidably in good manufacturing practice. It shall not contain added blood meal or SRM as defined in Section 6.1 of the Health of Animal Regulations. If it bears a name descriptive of kind, it shall correspond thereto. If one or more preservatives are used, the common name or names shall be indicated on the label. It shall be labelled with guarantees for minimum crude protein, minimum pepsin digestible protein (as determined by A.O.A.C. method 7.048, 13th edition), maximum moisture, maximum ash and minimum phosphorus. If the product contains "prohibited material" as set forth in Section 162(1) of the Health of Animals Regulations, it shall be labelled with the following statement(s) required by the Minister, in English and/or French: "Feeding this product to cattle, sheep, deer or other ruminants is illegal and Is subject to fines or other punishments under the Health of Animals Act"

SCHEDULE IV PART II	
CLASS 4. ENERGY FEEDS	
4.12	Palm fatty acid distillates, fractionated (or Fractionated palm fatty acid distillates or Fractionated PFAD) consists of the spray-dried solid fraction of palm fatty acid distillates, a by-product of edible palm oil production. It consists predominantly of saturated long chain free fatty acids, resulting from a fractionation process. It shall contain less than the upper-bound concentration of 0.75 nanograms WHO-TEQ per kilogram for dioxins and furans and less than the upper-bound concentration of 1.5 nanograms WHO-TEQ per kilogram for dioxins, furans and dioxin-like PCBs. This ingredient is for use as an energy source in ruminant feeds. If an antioxidant is used, it must be approved for use in livestock feeds, it shall be used at the approved rate, and the common name or names shall be indicated on the label. It shall be labelled with guarantees for minimum percent total fatty acids, maximum percent free fatty acids, maximum percent unsaturated fatty acids, maximum percent moisture, maximum percent unsaponifiable matter, and maximum percent insoluble matter.

Figure 1.21: Examples of approved ingredient descriptions from Schedule IV of Canada's Feed Regulations. Feeds containing only ingredients listed in Part I of Schedule IV are generally exempt from registration; feeds containing any of the ingredients listed in Part II must be registered.

Mineral supplements

Virtually all diets for every class of livestock require mineral supplementation. For lactating cows, supplementary minerals will normally be incorporated into mixed feeds containing other bulk dietary ingredients. For calves, heifers and dry cows, minerals may be included in a concentrate mix and/or may be provided as supplementary mineral mixes, available in one of several forms:

Complete mineral mixes contain blends of macro- and trace minerals, often with vitamins A, D and E added. Every feed dealer will sell a wide variety of these products, formulated for specific classes of livestock and particular geographic requirements. Commonly sold as 'licks' or in 10 or 25 kg bags, complete mineral mixes are designed to be offered to animals 'free-choice' (*ad libitum*) or to be blended with other grains and concentrates on-farm.

Unless there is no other practical choice, *ad libitum* feeding is not recommended except for iodized or cobalt-iodized salt. Although the rate of disappearance of mineral from an *ad libitum* feeder, block or 'lick' may seem appropriate for a group of animals, product loss from weather and spillage will almost always occur and individual intakes will vary widely. An often-heard argument in favour of *ad libitum* feeding of minerals is that animals have 'nutritional intelligence'—that they know what they need. This is no more true for cattle than it is for humans.

Macro premixes are specifically designed for addition to other concentrates in complete concentrate mixes used to supplement home-grown forages or in complete diets including forages (e.g., total or partial mixed rations). Containing both macro- and trace minerals as well as vitamins, these premixes are most commonly blended with other ingredients in the feed mill or on-farm.

Micro premixes are used in the same way as macro premixes but they contain only micro- (trace) minerals and vitamins and are, therefore, used in smaller quantities. When micro premixes are used, the macro-minerals are usually blended into the complete concentrate mix individually.

Supplementary minerals are most commonly added to diets in the form of inorganic salts (e.g., zinc sulfate, manganese carbonate, sodium selenite). Alternatively, minerals may be provided in the form of chelates—minerals complexed with organic molecules such as amino acids or proteins (referred to as proteinates). Claims are often made that chelated minerals are preferable because they are more bioavailable but the practical relevance of such claims is controversial.

Figure 1.22: Ad libitum feeding of minerals is not recommended except for iodized or cobalt-iodized salt.

Vitamin supplements

Supplementary vitamins are added to mixed feeds for all classes of dairy cattle. They may be included in a concentrate mix and/or may be provided with supplementary mineral mixes, as described above. Although most forages and grains contain substantial quantities of vitamins at harvest, by the time these feeds are consumed, vitamin levels may have dropped considerably due to the effects of sunlight, rain damage, heat, oxidation and mould growth.

Because rumen microbes can synthesize all of the essential B-vitamins, mixed feeds for ruminants usually provide only the fat-soluble vitamins: A, D and E. In most cases, these provide adequate intakes of A and D. However, unless specifically formulated for the requirements of high production (rapidly growing calves, cows in early lactation), the vitamin E levels in these mixes are generally inadequate, vitamin E being added primarily for the purpose of reducing oxidation of other ingredients.

Vitamins may also be delivered by intramuscular injection, as an alternative to feeding. The injection route is particularly appropriate for animals who are not consuming mixed concentrates, such as heifers fed primarily forage diets during winter.

Feed Additives

Feed additives are defined as non-nutritive substances added to feeds to:
· improve the efficiency of feed utilization;
· stimulate growth or other types of production;
· increase feed acceptance;
· enhance food safety and stability;
· improve the health or metabolism of the animal.

Medicating ingredients

In Canada, any feed additive that is classified as a 'medicating ingredient' (e.g., antibiotic, dewormer, ionophore, growth promotant) is regulated by the *Feeds Act* and *Regulations* which are administered by the Canadian Food Inspection Agency (CFIA). The CFIA's Compendium of Medicating Ingredient Brochures (CMIB) provides guidance on those medicating ingredients permitted by Canadian regulation to be added to livestock feed. This includes drug products that may only be used under a veterinarian prescription as well as products that may be used in the manufacture of livestock feed without veterinarian approval (over the counter products). Table 1.8 on page 18 lists medicating ingredients permitted in Canadian dairy diets as of October 2022.

The brochure for each ingredient describes the species and class of livestock it can be fed to, the level of medication permitted, directions for feeding and the purpose for which each medicating ingredient may legally be used, as well as approved brands and labeling requirements. As an example, figure 1.23 illustrates the brochure for the feed additive poloxalene.

Figure 1.23: Medicating ingredient brochure for the bloat prevention feed additive poloxalene.

As a complement to the CMIB, Health Canada's Drug Product Database (DPD) contains product-specific information on approximately 47,000 products that are currently approved for use in Canada, including both human pharmaceutical and biological drugs, veterinary drugs, radiopharmaceuticals and disinfectant products.

Non-medicating feed additives

Enzymes: A few research trials have demonstrated that fibrolytic (fibre-digesting) enzymes can, under some conditions, improve dry matter and neutral detergent fibre digestibility when added to lactation diets. These products have typically been produced as by-products of bacterial or fungal (e.g., yeast) culture and sold as 'microbial fermentation products' or 'microbial fermentation product extracts'.

Buffers: Buffers are salts of a weak acid or base that resist pH change, including Na, K, or Mg bicarbonate, and Ca carbonate. Buffers are fed to ruminants consuming diets containing high concentrations of fermentable carbohydrates to reduce ruminal acidosis. Sodium bicarbonate is typically fed with Mg oxide: Na bicarbonate: 110 – 220 g/d; Mg oxide: 45 – 90 g/d.

Probiotics: These are defined as direct-fed live microbials which, when fed to an animal, provide beneficial effects. They must be non-pathogenic, host-specific and viable in the digestive tract. Their normally short lifespan in the gut means that repeated dosing is required to maintain an effective population.

Animal class	Name of medicating ingredient	Status	Name of approved brand(s)	Primary claim (purpose)
All classes of cattle	Decoquinate	OTC	Deccox 6% Premix	Coccidiostat
All classes of cattle	Fenbendazole	OTC	Safe-Guard Premix 20%	Dewormer
All classes of cattle	Morantel tartrate	OTC	Banminth II 20% Premix	Dewormer
All classes of cattle	Poloxalene	OTC	Bloat-Guard	Bloat prevention
Dairy cows (dry and lactating)	Monensin	OTC	Coban Premix Rumensin Premix Monensin Premix Monovet Monensin Premix	Increased feed efficiency
Calves	Amprolium	OTC	Amprol Feed Premix	Coccidiostat
Calves (up to 136 kg including milk replacers)	Chlortetracycline hydrochloride	VP	Aureomycin 220 G Granular Medicated Premix Chlor 50 Chlortetracycline Premix Chlor 100 Granular Medicated Premix Deracin 22% Granular Premix	Diarrhoea prevention
Calves (including milk replacers)	Decoquinate	OTC	Deccox 6% Premix	Coccidiostat
Calves	Lasalocid sodium	OTC	Avatec 20 Lasalocid Sodium Premix Bovatec 20 Lasalocid Sodium Premix	Coccidiostat
Calves	Oxytetracycline hydrochloride	VP	Terramycin – 50 Terramycin – 100 Terramycin – 200 Oxysol-110 Oxytetracycline 50 Granular Premix Oxytetracycline 100 Granular Premix Oxytetracycline 200 Granular Premix	Bacterial enteritis prevention
Calves (milk replacers)	Decoquinate	OTC	Deccox-M Premix	Coccidiostat

Table 1.8: Medicating ingredients approved for use in dairy cattle feeds in Canada, as of January 2022. Status: OTC - Over the Counter; VP - Veterinary Prescription.

Potential modes of action of probiotics include:

· improvement of the intestinal microbial balance, excluding or reducing potentially pathogenic bacteria by competitive inhibition;
· synthesis of lactic acid with consequent reduction in intestinal pH;
· adhesion to intestinal mucosa, competitively reducing pathogen adhesion;
· stimulation of immune responses in the gut.

Most bacterial probiotics are based on *Lactobacillus acidophilus* or lactic acid-producing *Bifidobacteria*. Potential benefits include:

· reduction in diarrhea in calves;
· reduction in rumen pathogens (e.g., *E. coli*);
· increased milk yield in dairy cows.

Yeast probiotics, fed primarily to lactating dairy cows, are usually either *Saccharomyces cerevisiae* or *Aspergillus oryzae,* freeze-dried to preserve their viability and metabolic activity. Reported benefits include:

· increased feed intake;
· greater milk yields;
· altered milk composition;
· altered rumen fermentation resulting in improved fibre digestion and higher rumen pH;
· increased nitrogen retention (muscle gain).

It should be noted that yeast culture products sold as feed additives do not contain significant numbers of viable cells but have been reported to have benefits similar to those listed above for live cultures.

Antioxidants: Unsaturated fatty acids in feeds are subject to oxidative rancidity, where one of the pair of double bonds between adjacent carbons (see page 52) is replaced by bonds to a single oxygen atom. Fat oxidation results in feed discoloration and deterioration, decreased activity of the fat-soluble vitamins (A, D and E) and reduced feed palatability. Antioxidants added to feed to preserve flavour, odour and texture include the natural antioxidants:

· vitamin E (alpha-tocopherol);
· vitamin C (ascorbic acid);
· rosemary extract; and
· citric acid.

Synthetic antioxidants include:

· ethoxyquin (Santoquin®);
· butylated hydroxytoluene (BHT);
· butylated hydroxyanisole (BHA).

Mycotoxin binders: Binders decontaminate feeds by binding mycotoxins strongly enough to prevent toxic interactions with the animal consuming the feed, and to prevent mycotoxin absorption across the gastrointestinal tract. They include activated carbon, aluminosilicates (clay, bentonite), complex indigestible carbohydrates such as cellulose (for monogastrics), yeast cell wall components (glucomannans, peptidoglycans), and synthetic polymers.

Rumen-protected amino acids: As explained on page 57, two sources of amino acids are available to the ruminant animal for absorption from the small intestine: microbial protein and undegraded feed protein. If the relative concentrations of essential amino acids absorbed from these two sources does not match the animal's requirements, the excess amino acids may be wasted (see pages 49 and 82).

A number of studies have demonstrated production responses when lactating cows fed typical US corn grain / corn silage based rations received supplemental rumen-protected (RP) lysine and/or RP methionine, suggesting that these amino acids may be limiting among the essential amino acids absorbed. However, there is little evidence that similar responses might be expected in lactating cows fed typical western Canadian diets since the amino acid composition of undegraded feed proteins will be markedly different from that in diets commonly fed to US cows.

Anti-nutritive factors in feed

Various feeds may contain chemicals that are either toxic to the animals that consume them or interfere with normal digestive processes.

Ergot: Small grain crops are susceptible to infection by the fungus *Claviceps purpurea* whose wind-carried spores infect the developing flower and form a black-purple body (ergot body; sclerotia) in place of the kernel, as shown in the photo on page 10. Rye and triticale are most susceptible to infection.

Sclerotia contain toxic alkaloids that can produce a variety of symptoms in animals that consume them. In cattle, early signs of ergotism may include an elevated respiration rate, weight loss, reduced milk production, low conception rates and/or abortion. Continued consumption may lead to reddening, swelling, coldness, loss of hair, lack of sensation followed by development of blue-black color and dryness of the skin. Eventually, affected parts become separated and may slough off; gangrene caused by damage to capillaries in surface tissues and reduced blood flow can follow.

Gossypol is a natural pesticide produced by the cotton plant and is found in free and bound forms in whole cottonseed, cottonseed meal and cottonseed oil, any of which may be incorporated into dairy diets. Toxicity of the free form limits the use of these ingredients without treatment to lower free gossypol concentrations. The most common result of toxicity is the impairment of male and female reproduction; other clinical signs may include respiratory distress, impaired body weight gain, anorexia, weakness, apathy, and death after several days. Another important toxic effect of gossypol is its interference with immune function, reducing an animal's resistance to infections and impairing the efficacy of vaccines. Heat treatment of whole cottonseeds and cottonseed meal and refining of cottonseed oil are effective in reducing the risk of toxicity from free gossypol. However, digestion can release the bound form, resulting in increased concentration of the free form.

Trypsin inhibitors are proteins found in rye, to a lesser degree in barley and wheat, and in raw soybeans. They can form complexes with trypsin and chymotrypsin in the small intestine, inactivating these enzymes and thereby reducing protein digestibility in monogastric animals. Ruminants are not susceptible due to microbial digestion of the proteins but the pre-ruminant calf may be. The common practice of roasting soybeans inactivates the inhibitors.

Mycotoxins: Fungal infestation (mould growth) can occur at any time during plant growth, harvesting, storage or processing, primarily influenced by moisture level, temperature, and availability of oxygen. Not all moulds that grow on plants produce toxins and the mycotoxins that are produced by different moulds vary in their effects on the animals that consume them.

Table 1.9 provides a summary of the mycotoxins commonly found in animal feeds and their effects on animal health.

Nitrates: As described earlier, nitrate is the form of nitrogen the plant roots take up from the soil, from which it is transported to the leaves. When plants are stressed, excess nitrates may accumulate. Drought or hot, dry winds put forage under water stress, often resulting in nitrate accumulation. Damage caused by hail or frost impairs photosynthesis, resulting in excess nitrates. Cool, cloudy weather can also cause the problem. When any of these conditions occur within a few days before harvest or grazing, the potential for nitrate poisoning exists. If the stress is removed and the plants recover, nitrate levels will return to normal within several days.

In ruminant animals, nitrate is converted to nitrite and subsequently ammonia by rumen microbes. Nitrate poisoning occurs when the nitrite level in the rumen exceeds the capacity of the microbes to convert it to ammonia. When this happens, both nitrate and nitrite are absorbed into the

Mycotoxin	Effect on animals
Aflatoxin	liver disease, carcinogenic and teratogenic effects
Trichothecenes	immunologic effects, hematological changes, digestive disorders, edema
Zearalenone	estrogenic effects, atrophy of ovaries and testicles, abortion
Ochratoxin	nephrotoxicity, mild liver damage, immune suppression
Fumonisin	pulmonary edema, hepatotoxicity, leukoencephalomalacia, nephrotoxicity
Ergotamine	vasoconstriction, necrosis of the extremities, agalactia, abortion
Vomitoxin	feed refusal, poor weight gains; negative health effects are rare

Table 1.9: Mycotoxins commonly found in animal feeds and their effects on animal health.

bloodstream. Nitrite combines with hemoglobin to form methemoglobin, reducing the oxygen carrying capacity of the blood. When enough hemoglobin is converted to methemoglobin, the animal begins to suffer from oxygen deprivation.

Nitrate poisoning is much less of a concern for monogastric animals where nitrate is converted to nitrite in the large intestine, providing less opportunity for the nitrites to be absorbed into the bloodstream.

Dicoumarol: Apparent vitamin K deficiency may occur when mouldy hay is fed. In particular, sweet clover (figure 1.23) contains coumarin, the substance responsible for its sweet smell. Moulds on sweet clover hay can convert coumarin to dicoumarol

Figure 1.23: Sweet clover can be a source of dicoumarol, which reduces the activity of vitamin K.

which effectively reduces the activity of vitamin K in the blood clotting process, resulting in a risk of hemorrhage and death.

Listeria: Mouldy or spoiled silage that has not reached pH 4.0 – 4.5 during fermentation may harbour *Listeria monocytogenes*, bacteria that normally dwell in soil and in the gastrointestinal tract. Cattle consuming contaminated silage can develop Listeriosis, commonly referred to as 'circling disease' resulting from infection of the central nervous system. Contamination is most commonly due to soil incorporation during harvest.

Signs of infection may include depression, fever (40 – 41°C), weakness, incoordination, circling, nasal discharge, loss of appetite, or facial paralysis—a drooping eyelid or ear on one side. The neck may flex away from the affected side; the animal may lean or push on stationary objects. Pregnant animals may experience placentitis, fetal infection and death, abortion, stillbirths, neonatal deaths, and metritis.

Chapter 2: Feed Processing and Storage

Forage harvesting and storage

Forage crops, cut at an appropriate stage of maturity, are either allowed to dry for storage as hay (greenfeed when referring to grain crops) or immediately stored in the wet state as silage. The appropriate stage of crop maturity at harvest depends on the livestock to which the forage is to be fed. For example, alfalfa harvested at the early bud stage with a crude protein content of 16 - 18% is appropriate for lactating dairy cows but no class of beef animal requires a protein level that high. Harvesting at an early stage of maturity sacrifices crop yield in favour of higher feed quality as illustrated in figure 2.1.

Hay harvesting and storage

Hay is produced by field-drying green forage (legumes, grasses) to a moisture content of 15% or less. Ideally, mowing is done in the afternoon when crop sugar content is highest, due to the high rate of photosynthesis occurring at that time of day. When cut with a sickle-bar or flail mower, a wide, flat windrow (swath) is formed which, due to its large surface area, promotes drying but also risks significant damage from rain. The flat windrow may be subsequently raked to form a bulkier windrow and to promote drying by exposing the bottom of the swath.

The time taken to dry hay is weather-dependent. While the crop is drying, it will wilt while its soluble sugar content declines due to ongoing cellular respiration. In an effort to reduce these losses and prevent crop damage from rain, one or more strategies may be employed:

• the mower may be equipped with corrugated rollers that crimp or crush the stems after cutting (referred to as a mower-conditioner, figure 2.2), increasing exposure to air by breaking the stems' outer walls and increasing the aeration volume of the resulting windrow;

Figure 2.1: As forage crops mature, yield increases while nutritional value declines.

• after cutting, a 'tedder' (figure 2.3) may be used to 'fluff up' the windrow, providing more air exposure;
• a chemical dessicant may be sprayed on the crop at cutting to promote drying.

When the hay is sufficiently dry, it is picked up and either stored in loose form or baled in one of a number of shapes and sizes:

• small square bales, usually approximately 14 x 18 x 36 inches, weighing 50 - 80 pounds;
• large square bales, ranging from 36 - 48 inches in width and height, usually 96 inches long, weighing 800 - 1,500 pounds;
• large round bales, ranging from 48 - 60 inches wide, 48 - 72 inches in diameter, weighing 600 - 1,750 pounds.

The main objective in storing hay is to preserve its nutritional quality. Hay should be put into storage at a moisture level of no greater than 15%. Higher moisture levels may result in mould growth and heat damage. This is particularly true for large bales where the lower surface area to volume ratio reduces the rate of moisture and heat dissipation, compared with small square bales.

Figure 2.2: A mower-conditioner harvesting grass

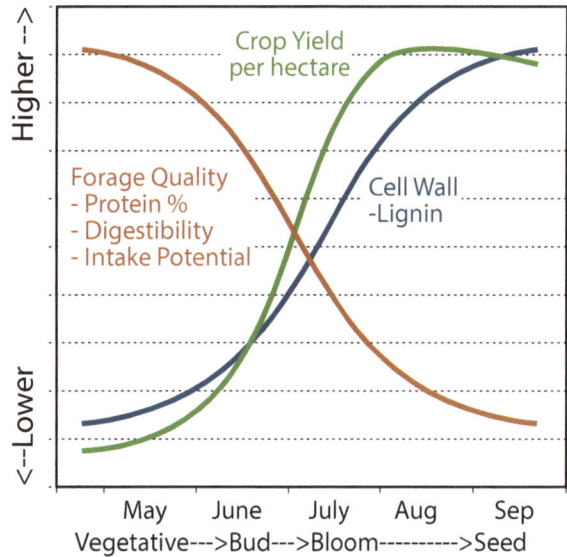

Figure 2.3: A tedder 'fluffing-up' a windrow

Figure 2.4: Baling large square bales.

Figure 2.5: Covered storage prevents weather damage

Hay stacks or individual large bales must be covered, by either a roof or a windproof tarpaulin, to minimize weather damage. The ground under the storage area must be well drained and sloped to prevent surface water runoff from entering the site.

| Stage | Forage Type | |
| | Legume | Grass |
	Dry Matter Losses (% of DM)	
Mowing	1–3	1–2
Mowing and Conditioning	1–4	1–2
Tedding	2–8	1–3
Swath Inversion	1–3	1–3
Raking	1–20	1–20
Baling		
Small Rectangular	2–6	2–6
Large Rectangular	1–4	1–4
Large Round	3–9	3–9
Hay Storage		
Inside	3–9	3–9
Outside	6–30	5–22

Table 2.1: Potential dry matter losses at each stage of hay harvest and storage. source: Alberta Forage Manual, Alberta Agriculture

Nutrients can be lost from hay during any stage of harvest or storage:

- cellular respiration: 4 - 15% of initial dry matter may be lost during wilting;
- legumes are particularly susceptible to leaf shattering and loss if the crop has to be raked or tedded, and during baling, especially if the crop is very dry;
- rain damage can leach nutrients from the crop as it lies in the field or in storage if it is not covered;
- vitamins, particularly vitamin A may be lost due to UV exposure in the field;
- if moisture content exceeds 15 - 20%, heat produced by cellular respiration and/or microbial activity can result in the formation of heat-damaged protein (see ADF nitrogen in Feed Analysis section, page 33), mould growth and spontaneous combustion.

The ranges of losses possible at each stage are illustrated in table 2.1.

Silage harvesting and storage

Strategies used to promote hay drying are not required when cutting crops to be stored as silage. When the crop is to be stored in a bunker or tower silo or in an Ag-Bag® type of system, it is typically cut, windrowed (swathed) and allowed to wilt for a few hours to achieve a moisture content of 30 - 40%. The windrow is then picked up and chopped directly into a forage wagon, as illustrated in figure 2.7, and moved immediately into storage.

For corn silage, swathing and wilting are not required when the crop is harvested at the optimum moisture level: 50 - 60% for upright oxygen-limiting silos; 60 - 65% for upright stave silos; 60 - 70% for bags; 65 - 70% for bunkers.

Figure 2.6: Corn silage is typically cut and chopped directly into the forage wagon for immediate ensiling.

Another alternative is putting up silage in individual large round bales. In this case, the crop is cut and windrowed, allowed to wilt, then directly picked up and baled. The silage bale is then wrapped in plastic, either before being placed on the ground or after being picked up by the wrapping machine in a subsequent operation.

Figure 2.7: Harvesting silage after wilting.

Figure 2.8: Filling and packing a bunker silo.

Figure 2.9: Stave (left) and oxygen-limiting (right) upright silos.

Figure 2.10: Round bale silage being wrapped in plastic.

No matter how the silage is to be stored, moisture content is critical. If the moisture level in the crop is too low, exclusion of oxygen from the silage mass is difficult, allowing heating and mould growth. If the crop is too wet, seepage from the silo will result in a loss of nutrients and clostridial fermentation may predominate, producing large amounts of malodorous butyric acid, lowering palatability.

Figure 2.11: 'Ag-Bag®'-type silage storage.

The keys to making good silage are to:

- harvest the crop from the field at optimum maturity to achieve the appropriate nutritive value;
- wilt the crop as quickly as possible to an acceptable dry matter content (30 – 40% for grasses; 35 – 40% for legumes) to preserve nutrients and promote a good fermentation;
- chop forage particles to an optimal size to allow for adequate fibre length and effective packing;
- rapidly exclude air from the forage mass in the silo;
- seal the silo as rapidly as possible after filling;
- prevent the penetration of air into the silage mass during storage and subsequent feed out.

An initial stage of aerobic respiration will consume most of the oxygen remaining in the silage, followed by a prolonged phase of anaerobic fermentation by a variety of lactic acid-producing bacteria. The pH level of the forage will be between 5.5 and 6.0 prior to ensiling, ultimately reaching pH 4 – 4.5. This low pH essentially pickles the forage and, coupled with the anaerobic environment, prevents the growth of spoilage microbes like clostridia, yeasts and moulds.

Bacterial inoculants are often added to silage as it is placed in the silo to improve the probability of a successful fermentation phase. Inoculants typically contain fast growing homo-fermentative lactic acid bacteria (producing only lactic acid) that dominate the fermentation, increasing the rate of fermentation, rapidly lowering the pH and assuring the anaerobic stability of the silage. Compared to untreated silages, those treated with homolactic acid bacteria are often lower in pH, acetic acid, butyric acid and ammonia-N (a product of protein breakdown) but higher in lactic acid content.

Dehydrated alfalfa products

Several companies in western Canada produce high quality dehydrated alfalfa pellets and 'cubes' (commonly referred to as 'dehy') for animal feeding. To minimize loss of nutrients, alfalfa is partially wilted in the field, chopped green and then dried in rotating drums, retaining its bright green colour and vitamin A activity.

Figure 2.12: Dehydrated alfalfa pellets (top - 0.5 in. diam.); dehydrated alfalfa 'cubes' (bottom - 1.25 in. on 2 sides).

Grain and concentrate processing

The fibrous outer hulls of most whole grains reduce digestibility by restricting access to the kernel by microbes and digestive enzymes. Therefore, they are usually milled or ground to improve digestibility and to facilitate uniform mixing with other ingredients.

Roller milling

In a roller mill, whole grain is fed through a pair of closely spaced cylindrical rollers rotating in opposite directions, breaking the hull and flattening the kernel. Roller spacing is adjustable and the roller surfaces may be either smooth or corrugated. If the grain is dry, rolling will break the hull and kernel into several pieces, facilitating blending with other ingredients but often producing excessive fines (small particles).

By steaming the grain for 10 - 15 minutes to increase grain moisture to 12 - 14% before processing, *steam rolling* reduces the production of fines, resulting in a thick flake. In the *steam flaking* process, grain is steamed for 30 - 60 minutes in a vertical steam-chamber to increase grain moisture to 18 - 20% and then rolled between preheated rollers to drive off excess moisture. A number of research trials have demonstrated improved digestibility of steam-flaked grain compared with dry rolled or steam-rolled grain, likely due to the expansion of the kernel, facilitating greater access to starch and other components by microbes and digestive enzymes.

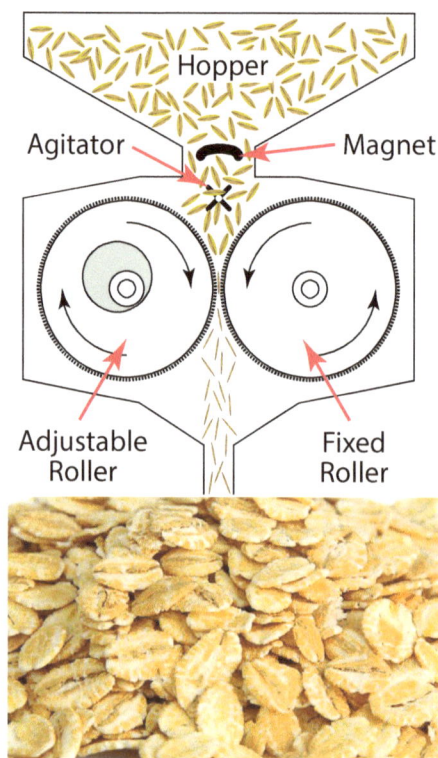

Figure 2.13: Schematic of a typical roller mill used to produce steam-flaked barley.

Temper rolling is an alternative to steam rolling that can be used on-farm—whole grain is rolled after soaking overnight in water. This process reduces shattering of the grain, resulting in fewer small starch particles which are very rapidly degraded by rumen microbes, increasing the risk of ruminal acidosis.

Hammer milling

If the primary objective is to reduce the particle size of grain to facilitate mixing with other ingredients, hammer milling is often the logical choice. Rotating 'hammers' shatter the dry grain as it enters the grinding chamber. The size of the resulting particles depends on the size of the perforations in the grate through which particles must pass.

Dry rolled and milled grains are often mixed with other dietary ingredients and offered to livestock in 'mash' form. The disadvantage of this method of feeding is that ingredients of varying sizes and densities are likely to separate in transport, storage and feeding. When fed to groups of animals, it is impossible to be assured that each individual is consuming the desired proportion of each ingredient.

Pelleting

Pelleting prevents separation of ingredients, ensuring that all animals consume the prescribed proportions of each. The process of pelleting consists of forcing steam-softened single or multiple mixed ingredients through holes in a metal die plate to form compacted pellets which are then cut to a pre-determined length. In addition to softening the feed, the steam partially gelatinizes the starch content of the ingredients, resulting in firmer (and for aquafeeds, more water-stable) pellets. Pellet binder ingredients are often included to prevent pellets from crumbling during handling.

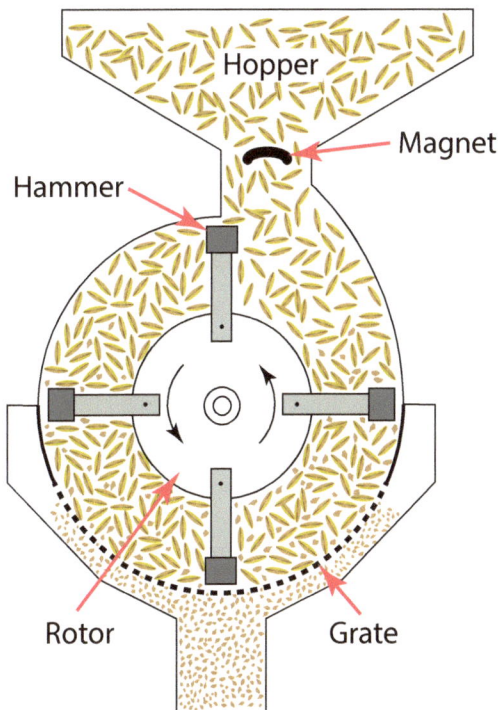

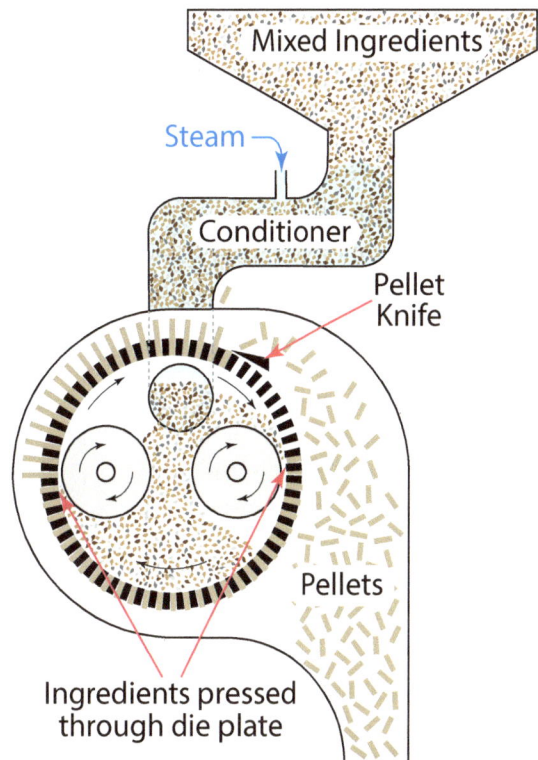

Figure 2.14: Schematic of a typical hammer mill used to produce ground barley.

Figure 2.15: Schematic of a typical pellet mill used to produce the type of pellets shown.

Extrusion

In the extrusion process, mixed ingredients are steam-cooked under high pressure and temperature (80 – 200°C) producing a high level of starch gelatinization. The resulting dough is then forced through a die by a screw press to produce pellets of a desired shape. Extruded pellets are very stable in water, making them very desirable as aquafeeds. After cooling, fat or another palatability enhancer is sprayed on and the pellets are dried to less than 10% moisture content. Like steam flaking, extrusion increases the digestibility and palatability of the feed.

Extrusion is the most common method of preparing pet foods sold in the form of 'kibble' which comprise over 60% of the market for dog and cat food in North America. It is not commonly used to produce ingredients for dairy diets except for calf starters.

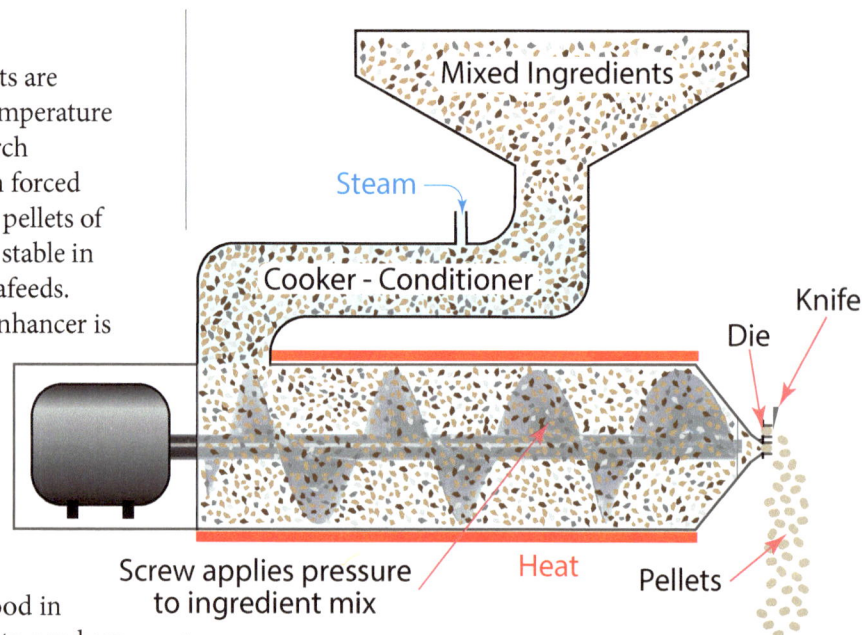

Figure 2.16: Schematic of a typical extruder used to produce the type of pellets shown.

Chapter 3: Feed Sampling and Analysis

Before describing the methods used to analyze feed, it should be recognized that dairy cattle diets seldom consist of a single feed; they are normally formulated by combining several ingredients, either fed separately or mixed together in various proportions. Each ingredient contributes its constituent nutrients to the total diet (see Chapter 7: Ration Formulation).

A second consideration is that only a small sample of a larger quantity of each feed ingredient is normally analyzed. Therefore, obtaining a sample for analysis that is representative of that larger quantity is essential. Specific tools and protocols (described below) are available for obtaining representative samples from the various types of common dietary ingredients.

Finally, it should be recognized that feed samples are seldom analyzed for the pure organic (carbon-containing) nutrients they contain, such as individual amino acids or fatty acids (see Chapter 5: Nutrients). Routine feed analyses generally only assess the contribution of broad classes of feed fractions, such as crude protein, crude fat or structural carbohydrates.

Feed analysis is aimed at estimating potential contributions to dietary requirements. The common analyses have their limitations and a brief description of the methods used will provide an understanding of how the results can be used. A typical commercial feed tag is shown in figure 3.1; a typical feed lab forage analysis report is shown in figure 3.7 (page 35). Examples of standard 'book' values, commonly used in diet formulation, are illustrated in Appendix A.

Many of the forages grown in western Canada are of insufficient quality to meet the minimum requirements of dairy livestock without supplementation. For example, a significant proportion of our grass hays have crude protein (CP) levels in the 7 – 10% range while 11% CP is considered minimal for dry cow maintenance.

Unless the nutrient content of dietary ingredients is known, their allocation to meet productive requirements is pure guesswork. Undernutrition is detrimental to productivity while overnutrition is a waste of valuable feed resources. Maximum dollar returns demand that animal requirements be matched by nutrient intake. This can only be accomplished when nutrient-contributing feed fractions have been determined through feed analysis.

Feed sampling

Feed testing involves both sampling and analysis. Often the importance of the former is underestimated. A feed testing laboratory can only analyze what is submitted and unless the sample is representative of the available feed supply, the time and money spent will be wasted.

18% Calf Starter

For prevention of coccidiosis caused by Eimeria bovis and Eimeria zuernii in ruminating and non-ruminating calves, including veal calves.

Active Drug Ingredient:
Decoquinate...50 mg/kg

Guaranteed Analysis:
Crude Protein, Min.. 18.0%
Crude Fat, Min.. 2.5%
Crude Fiber, Max 6%
Acid Detergent Fiber (ADF), Max........................... 10%
Calcium, Min 0.7%
Calcium, Max.. 1.2%
Salt, Min 0.4%
Salt, Max.. 0.8%
Phosphorous, Min 0.5%
Selenium, Min ....0.3 µg/g
Vitamin A, Min ....25,000 IU/kg

Ingredients:
Grain products, Processed grain by-products, Soybean meal, Linseed meal, Cane molasses, Monocalcium phosphate, Dicalcium phosphate, Calcium carbonate, Salt, Vegetable oil, Vitamin A acetate, Cholecalciferol (source of Vitamin D3), Vitamin E supplement, Calcium iodate, Magnesium oxide, Manganous oxide, Ferrous sulfate, Potassium chloride, Zinc oxide, Cobalt carbonate, Sodium selenite, Natural and artificial flavors, Ethoxyquin and BHT (preservatives).

Feeding Directions:
Feed 18% Calf Starter at a daily rate of 1 kg per 100 kg of bodyweight. This will provide 0.5 mg of decoquinate per kg of bodyweight. Feed at least 20 days during periods of coccidiosis exposure or when experience indicates that coccidiosis is likely to be a hazard. Coccidiostats are not indicated for use in adult animals due to continuous previous exposure to coccidia. Warning: Do not feed to cows producing milk.

Manufactured by: Dairy Feed Co., Nosehair SK

Net Wt. 25 kg

Figure 3.1: Example of a commercial feed tag.

For example, a bale is thrown from the top of a haystack, the strings are cut and a handful of hay is pulled from the centre, put in a plastic bag and sent in for analysis. Several sources of sampling error are possible:

- Perhaps the bale, being on top of the stack, was one of the last baled. Did that part of the swath receive more rain than that baled earlier? Had it dried more, resulting in greater leaf shattering? Was it from a corner of a field not typical of the rest?
- As a result of being on top of the stack was the sample bale more weathered? Had the stack received rain?

- In pulling hay from the bale, were leaves stripped off resulting in a sample with a high proportion of stems? The goal of feed sampling is to obtain a sample that is representative of the average feed quality of the bulk of feed from which it is taken. If there is reason to believe that significant differences exist between one batch of feed and another, then representative samples should be drawn from each batch separately.

Sampling hay

A convenient coring tool (figure 3.2) is available for sampling hay. A stack of bales should be sampled in 10 to 15 different locations, well distributed around the stack. Additional cores will result in a more representative sample. The cores should be thoroughly mixed in a plastic bucket and from this a sub-sample should be taken for submission.

The ideal method for obtaining a sub-sample is to dump the combined cores after mixing onto a flat, smooth surface to form a uniform pile. Then, using a large flat blade, divide the pile into quarters. If a quarter is too large to submit, use the blade to divide the quarter into halves or quarters again, until sub-samples of appropriate size are obtained. A second sub-sample should be saved in case a question arises out of the analysis results.

Sampling silage

Silage in bunkers, piles or bags can be sampled using an extended coring tool such as that shown in figure 3.3. If an appropriate probe is not available, or for silage stored in upright silos, the sampling strategies described in the text box on page 31 are recommended.

Sampling concentrates

Dry concentrates stored in bulk can be sampled by plunging an arm into the feed and sampling at least 10 different sites. If the feed is bagged, samples should be taken from five to ten individual bags, ideally using a sampling tool such as one of those shown in figure 3.4.

Figure 3.3: An extended probe used for sampling silage.

Use the following technique to get a homogeneous sample of grain or bulk concentrate as it is being unloaded. You will need four 20 litre pails (label two of them A and B), a cup and bags to retain 1 kg samples.

1. As each truck is unloaded: use the cup to take samples, from the centre and sides of the stream, every 30 - 60 seconds;
2. Place samples into pail A so that it is ¾ full by the time the truck is empty;
3. Thoroughly hand mix contents of pail A;
4. Place the two unlabeled pails side by side and pour contents of pail A in to the centre where these buckets meet, so that each pail receives half of the contents of A;

Figure 3.2: A coring tool used for sampling hay.

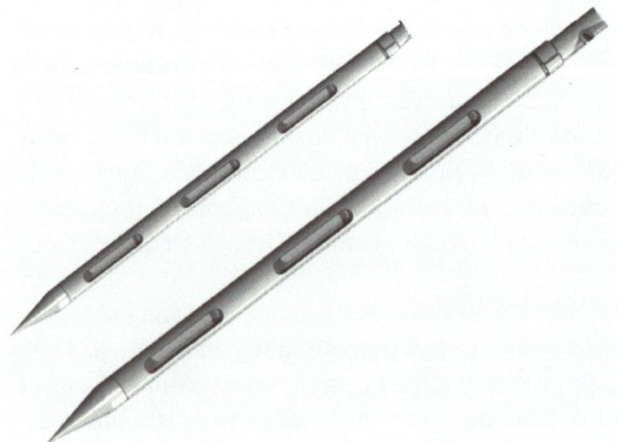

Figure 3.4: Bagged concentrate sampling probes (triers).

Silage sampling strategies

Silage samples should be obtained both as the silo is being filled and after the silage has been allowed to reach its final stage of fermentation:

- sample(s) obtained when filling the silo will give a good indication of forage quality early enough to properly plan the feeding program;
- samples taken when the silo is opened up will provide additional information on the success of the ensiling process and confirm the pre-ensiling analysis results.

Samples of wet, fresh forage pre-ensiling should be taken as follows:

1. Have a clean 5 gallon pail with a sealable lid located at the silo.
2. The person filling the silo should take a handful of fresh forage from every 4 loads and place them in the pail. Put the lid on the pail after each sample is added to retain the moisture.
3. At the end of each day, mix the fresh forage in the pail by hand and, taking a single handful of the forage, put it in a sealed bag in a freezer. Repeat this process each day for up to 7 days.
4. When finished filling that silo or those silo bags, add the samples together and submit for analysis. Refreeze the samples prior to shipping if necessary.
5. Separate samples should be taken and submitted if filling silo or silo bags takes longer than 7 days or if field or crop conditions change, for example, if a variety of silage is later maturing and therefore greener and wetter than other fields.

Samples of fully-fermented silage from a silo or bag should be taken as follows:

Pit or bunker silos:

1. Don't take the sample until the silo has been opened up and used for at least a week.
2. Take the sample at some point in the day after some silage has been removed. Avoid collecting any silage that has spoiled or has remained exposed for a day or more.
3. Take 20 handfuls of silage from all across the face of the silo going as high as you can reach. Put these into a clean pail and mix thoroughly.
4. Put a sub-sample (2 - 3 handfuls) of the silage collected in the pail into a plastic bag, freeze and submit to the lab.

Silage bags:

1. Don't take the sample until the bag has been opened up and used for at least 2 days.
2. Take 5 handfuls of silage from the face of the bag and place in a plastic bag in a freezer. Repeat for 5 days.
3. At the end of the 5 days, combine the 5 samples and mix thoroughly. Put a sub-sample (2 – 3 handfuls) in a bag and submit to the lab. Refreeze if necessary.

Tower silos:

1. Don't take the sample until the silo bag has been opened up and used for at least 5 days.
2. Take 5 handfuls of silage while the silo is unloading and place in a plastic bag in a freezer. Repeat for 5 days.
3. At the end of the 5 days combine the samples and mix thoroughly. Submit a sub-sample (2 – 3 handfuls) to a lab for testing. Refreeze if necessary.

NOTE: preserving the actual moisture content in the fresh forage or silage sample is very important. Always double bag silage samples. Freezing or refrigeration and exclusion of air from the sample are necessary to prevent spoilage during transportation to the lab. Send samples to the lab early in the week (Monday preferred) so that they don't sit over a weekend in a warm receiving area of the lab.

Packaging and labeling:

Label the sample clearly with an identification description, method or system. You must be able to locate the feed when the results come back if you are to realize the value of the feed test. For example:

Farm	Feed	Cut/variety	Date Collected
Belly Acres	Corn Silage	Maxim	29 Sept 2022
Footrot Flats	Alfalfa Silage	1st/Regal	5 Jul 2022

source: Dale Engstrom

5. Place the contents of one of the unlabeled pails into the bin and repeat step 4 with the remaining pail until you have a 2 kg sample to be placed in pail B;
6. Thoroughly hand mix the contents of pail B;
7. Remove a 1 kg sample from pail B bag and submit for analysis.

Nutrient concentrations in grains are much less variable than those in forages. In many cases, such as when grain is purchased frequently in small quantities, 'book values' are used. Examples can be found in Appendix A. On the other hand, nutrient levels in screenings and other by-products are impossible to predict. They should be routinely sampled as described above for concentrates and submitted for analysis.

Feed analysis

Chemical analysis of feed is aimed at estimating its potential to support animal health, maintenance and production. A brief description of the methods used will provide an understanding of how the results can be used to best advantage. Reference to the feed analysis report shown in figure 3.7 (page 35) will assist in putting the discussion into perspective.

Dry matter (DM) and moisture (line 1)

The amount of moisture contained in feeds is widely variable. Hay and grain usually contain about 10% moisture. Silage may contain 50 – 75%. Pasture plants are often 80 – 85% water. In most (but not all) feeding situations, animal intake is limited only by the dry matter content of a feed. In other words, a heifer capable of consuming 5 kg (11 lbs) of leafy grass hay (10% moisture, 90% dry matter) will also be capable of consuming 22.5 kg (~50 lbs) of leafy grass pasture (80% moisture, 20% dry matter). In both cases she will consume 4.5 kg (~10 lbs) of dry matter. Expressing feed analysis, animal intake and nutrient requirements on a dry matter basis eliminates moisture as a variable in the comparison of different feeds and in the formulation of balanced rations.

In the feed lab, the dry matter content of a feed sample is determined by drying to constant weight in a forced air oven at either:
- 100 to 105°C overnight;
- 135°C for 2 hours, or;
- 55 to 60°C for 24 hours followed by drying at 105°C for 48 hours.

This last method results in less browning and sample destruction than occurs when drying at the higher temperatures without the initial 55 to 60°C treatment.

On the feed analysis report, dry matter is expressed as a percentage. The figure is derived by simply weighing a sample of feed before and after drying:

$$DM\% = \frac{\text{dry weight}}{\text{wet weight}} \times 100$$

In addition to being used to correct feeds to the 100% DM basis, DM% also provides information about the storage properties of feed. Extra moisture can result in heating and spoilage in hay and grain. Inadequate moisture in silage may result in poor preservation while a high moisture content may lead to excessive nutrient leaching. However, for these purposes, dry matter content should be measured before the feed is stored, commonly done using an on-farm microwave oven (see sidebar on page 40). Analysis results from the feed lab are usually received too late for remedial action to be taken.

Guaranteed analysis values on commercial feed labels are stated on an 'as-fed' (AF) basis. These can be converted to the dry matter basis as follows:

$$DM \text{ value} = \frac{\text{AF value}}{DM\%} \times 100$$

Acid detergent fibre (ADF; line 2)

For the determination of acid detergent fibre (ADF), feed samples are boiled in a solution containing sulfuric acid and the detergent, cetyl trimethyl ammonium bromide. Hemicelluloses and cell wall proteins are dissolved, with the residue containing cellulose, lignin, lignified nitrogen, cutin, silica and some pectins. ADF% is simply the weight of the residue expressed as a percentage of the original sample. As described in the Energy section below (page 37), ADF% is commonly used to estimate the potential of feed to provide energy.

Neutral detergent fibre (NDF; line 2)

Neutral detergent fibre (NDF) is determined by boiling a sample of feed in a solution containing the detergent sodium lauryl sulfate, sodium sulfite to solubilize protein, and α-amylase to digest starch. This solution extracts lipids, sugars, organic acids and other water-soluble components as well as pectin, non-protein nitrogen (NPN) compounds, soluble protein and some of the silica and tannins. NDF is the insoluble residue made up of cellulose, hemicellulose, lignin, lignified nitrogen, some protein, minerals and cutin. NDF% is the weight of the residue expressed as a percentage of the original sample. Since it provides the most complete measure of cell wall components, NDF is used to balance fibre requirements in ration formulation.

Crude fibre (CF; line 3)

Crude fibre is one of the analytical fractions measured in the proximate system of analysis described on the page opposite. Although seldom used today in diet formulation, commercial feed labels are still required by law to indicate the maximum crude fibre content of the product (see figure 3.6 on page 34).

The residue remaining after extraction of crude fat from a feed sample of weight A (see method below) is sequentially boiled in sulfuric acid and sodium hydroxide. The remaining residue is washed with hydrochloric acid, followed by petroleum ether, dried at 105°C and weighed (weight B). This residue is burned in a furnace at 550°C, and the resulting ash is weighed (weight C).

$$\text{Crude Fibre } \% = 100 \times \frac{B - C}{A}$$

Crude fat (EE; line 3)

Lipids (fats and oils) are reported in the proximate analysis scheme as crude fat, commonly referred to as ether extract (EE).

A dried sample of feed is placed in a porous extraction thimble in a Soxhlet extractor apparatus (figure 3.5) and subjected to a continuous extraction with diethyl ether at 40–60°C for a defined period of time. After evaporation of the solvent, the residue will contain lipids, cutin, waxes, organic acids, alcohol and pigments.

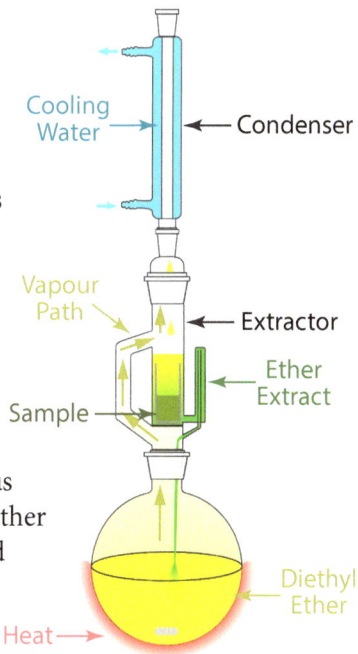

Figure 3.5: Soxhlet extractor used for crude fat analysis.

After extraction, a sample of the defatted feed residue remaining in the thimble is used for crude fibre analysis, as described previously (page 32).

Crude protein (CP; line 4)

Determined using the Kjeldahl procedure, crude protein is another of the fractions in the proximate analysis scheme described below. A dried sample is first digested in concentrated sulfuric acid with a mercury or selenium catalyst, which converts most of the nitrogen (N) to ammonium sulfate (N present as nitrate is only partially converted). This mixture is cooled, diluted with water and neutralized using sodium hydroxide, resulting in the dissociation of the ammonium sulfate. Distillation drives off ammonia and the distillate is titrated with acid to determine its ammonium concentration, from which the N level in the original sample is calculated.

Since most feed proteins contain about 16% N (ranging from 15.7 to 19.3%), CP% is estimated by multiplying the N concentration in the feed by 6.25—the inverse of 16% ($1 \div 0.16 = 6.25$). However, some portion of the N in most feeds is found as non-protein nitrogen (NPN) (see below) and, therefore, the value calculated by multiplying N x 6.25 is referred to as crude rather than true protein.

Non-protein nitrogen (NPN; line 4)

As noted above, crude protein is composed of two nitrogen-containing feed fractions: true protein and non-protein nitrogen (NPN). To separate these two fractions, true protein (TP; line 5) is precipitated out of solution using either tungstic acid or trichloroacetic acid, then dried and weighed. The NPN left in solution is measured using the Kjeldahl procedure as described for the determination of CP.

NPN, derived from compounds such as urea, ammonium salts, amino acids, small peptides and nucleic acids, represents the fraction of crude protein most rapidly degradable by microbes in the rumen.

Soluble crude protein (SCP; line 5)

The soluble crude protein (SCP) content of a feed is estimated by mixing a sample in borate-phosphate buffer. Crude protein (CP) which is soluble in the buffer is measured using the Kjeldahl procedure. SCP is an alternative to NPN to estimate the amount of CP which is rapidly degradable by rumen microbes.

ADF nitrogen (ADF-N; line 6)

When silage is put up too dry (greater than 50% dry matter) or hay too wet (less than 85% dry matter), excessive heating may cause some amino acids in crop proteins to become bound to certain simple sugars, creating indigestible complexes. Heating during feed processing (e.g., pelleting) can have the same effect.

The severity of heat damage is estimated in the feed lab as ADF nitrogen by measuring the amount

---

The Proximate System of Analysis

The proximate system for routine analysis of animal feedstuffs was devised in the mid-nineteenth century at the Weende Experiment Station in Germany. It was developed to provide a very broad classification of feed components. The system consists of the analytical determinations of dry matter, crude protein, crude fat (ether extract), crude fibre and ash. Nitrogen-free extract (NFE), more or less representing sugars and starches, is calculated by difference rather than measured by analysis:

% NFE = 100% – (CF% + CP% + EE% + Ash%)

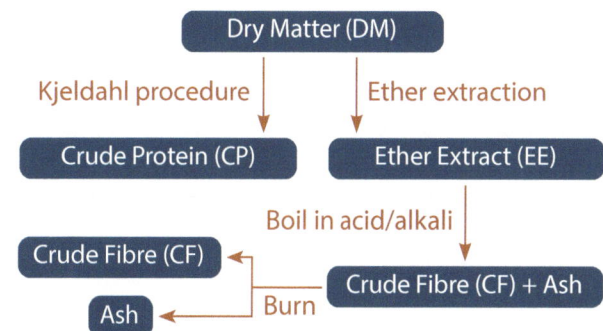

34% DAIRY SUPPLEMENT (PELLET) FOR LACTATING DAIRY COWS
GUARANTEED ANALYSIS

Crude Protein (minimum) 34.0%
Crude Fat (minimum)..1.1%
Crude Fibre (maximum) 4.8%
Calcium (actual) 2.8%
Phosphorus (actual) 1.85%
Sodium (actual)... 0.65%
Magnesium (actual)..1.2%
Iodine (actual)... 6.0 mg/kg
Copper (actual)...135 mg/kg
Manganese (actual) ....400 mg/kg
Zinc (actual)..410 mg/kg
Cobalt (actual) .... 1.0 mg/kg
Fluorine (maximum)..130 mg/kg
Vitamin A (minimum)................................. 69,000 IU/kg
Vitamin D (minimum) .... 11,300 IU/kg
Vitamin E (minimum) .... 270 IU/kg

Figure 3.6: Feed labels for manufactured complete feeds as well as supplements and macro premixes must include crude protein, crude fat and crude fibre guarantees.

of nitrogen (N) associated with the acid detergent fibre (ADF) residue using the Kjeldahl method. ADF Nitrogen may be reported as:
- acid detergent insoluble nitrogen (ADIN);
- acid detergent insoluble protein (ADIP);
- acid detergent fibre protein (ADF-P), or;
- heat-damaged protein;

expressed as a percentage of either total N, total crude protein or total dry matter. Nitrogen values are multiplied by 6.25 to convert to protein values.

Although ADF-N is widely accepted as a good measure of heat damage in forages, it may not be appropriate for use in non-forage protein sources such as dried distillers grains (DDG) where there may be very little correlation between ADIN and protein digestibility. In fact, with some non-forage protein supplements, heating may increase their value as rumen undegradable protein (RUP; see page 38).

NDF nitrogen (NDF-N; line 6)

NDF nitrogen represents nitrogen associated with the cell wall, measured by subjecting the neutral detergent fibre (NDF) residue to the Kjeldahl procedure. NDF nitrogen may be reported as:
- neutral detergent insoluble nitrogen (NDIN);
- neutral detergent insoluble protein (NDIP), or;
- neutral detergent fibre protein (NDF-P);

expressed as a percentage of either total N, total crude protein or total dry matter.

A fraction of NDF-N will be both degradable by rumen microbes and digestible in the small intestine. A second fraction will be completely indigestible. It is generally assumed that the completely indigestible fraction is estimated as ADF Nitrogen x 6.25. The degradable/digestible fraction is, therefore, calculated by difference: (NDF-N – ADF-N) x 6.25.

Starch (line 7)

Starch content is measured by first treating the sample with heat and moisture to gelatinize the starch. After cleaving its long chains of glucose molecules with enzymes, the glucose is treated with a second enzyme linked to a colorimetric indicator whose depth of colour correlates with starch concentration.

Ash (line 7)

Ash is the residue remaining after complete combustion of the crude fibre + ash fraction in the crude fibre method described earlier. Ash represents the total mineral content of the feed.

Minerals (lines 8-13)

As stated earlier, most of the minerals found in plants are associated with organic compounds. Mineral analysis involves complete combustion of a sample of the feed, leaving an ash in which the minerals are present as simple inorganic (no carbon) salts. The ash is subsequently analyzed for each element specifically.

Mineral content may be determined by any of a number of alternative methods, including atomic absorption spectrometry, atomic emission spectrometry, mass spectrometry, neutron activation analysis, x-ray emission spectrometry, molecular light absorption spectrometry, molecular fluorometry, electrochemistry, combustion elemental analysis, volumetry, ion chromatography or gravimetry.

Notice on the feed analysis report (page 35) that calcium, phosphorus, magnesium, potassium, sodium and chloride are stated as percentage values while iron, manganese, zinc, copper and molybdenum are reported as mg/kg (milligrams per kilogram) and selenium as µg/kg (micrograms per kilogram). This is simply a reflection of the fact that some minerals are found in plants in relatively large amounts while others are present in only trace quantities. Some feed labs will report trace mineral levels in units of ppm or ppb:

1% = 1 part per hundred
1 µg/g = 1 mg/kg = 1 ppm (part per million)
1 µg/kg = 1 ppb (part per billion)

AgriLabs
International Inc.
Feed · Soil · Water

#167, 386 Lower Main Street
Bigtown BC V3W 4X6 Canada
(250)542-6111 · fax: (250) 542-6112
info@agrilabs.ca · www.agrilabs.ca

Submitted by:

Joe Milkman
Com 1 Site 3 RR#2
Upper Cutbank BC V2Z 0X0

Feed Analysis Report

Date received: 17 Aug 2022
Date reported: 21 Aug 2022
Sample retained: 28 Aug 2022
Sample received via: Courier

Lot number: 123456
Report number: 789101
Sample ID: 12131415
Page: 1 of 1

Sample description:

second cut
bromegrass hay
north quarter
baled 5 August 2022

L	Analysis	Units	Result	Analysis	Units	Result
1	Dry Matter	%	88.3	Moisture	%	11.7
2	Acid Detergent Fibre	% DM	40.3	Neutral Detergent Fibre	% DM	65.9
3	Crude Fibre	% DM	34.6	Crude Fat (EE)	% DM	1.64
4	Crude Protein	% DM	8.34	Non-protein Nitrogen	% DM	1.12
5	True Protein	% DM	7.22	Soluble Crude Protein	% CP	24.6
6	ADF Nitrogen	% DM	0.23	NDF Nitrogen	% DM	0.73
7	Starch	% DM	2.64	Ash	% DM	8.84
8	Calcium	% DM	0.55	Phosphorus	% DM	0.18
9	Magnesium	% DM	0.18	Potassium	% DM	1.59
10	Sodium	% DM	0.04	Chloride	% DM	0.19
11	Iron	mg/kg	147	Manganese	mg/kg	67.4
12	Zinc	mg/kg	16.2	Copper	mg/kg	5.96
13	Selenium	ug/kg	116	Molybdenum	mg/kg	2.23

L	Calculated Value	Units	Result	Calculated Value	Units	Result
14	Non-fibre Carbohydrates	% DM	15.3	Non-structural Carbohyd	% DM	19.8
15	Adjusted Crude Protein	% DM	8.34	Relative Feed Value	%	81.1
16	Digestible Protein	% DM	4.51	Metabolizable Protein	% DM	3.15
17	Rumen Degradable Prot	% CP	65	Rumen Undegradable Prot	% CP	35
18	Total Dig Nutrients	% DM	58	Digestible Energy	Mcal/kg	2.55
19	Metabolizable Energy	Mcal/kg	2.09	Net Energy, maintenance	Mcal/kg	1.27
20	Net Energy, gain	Mcal/kg	0.73	Net Energy, lactation	Mcal/kg	1.27

Figure 3.7: Example feed analysis report.

Later, in the description of nutrient requirements, minerals will be divided into macromineral and trace or micromineral categories. This is not meant to imply that the minerals required in trace quantities are any less important than those required in larger quantities.

Mineral levels given on the feed analysis report cannot be considered in isolation, since many of the elements interact with one another, affecting availability (see page 61). For example, a copper level of 8 μg/kg might appear adequate with reference to the requirement tables in Appendix A. However, if the molybdenum level is more than 2 μg/g, the availability of copper to the animal may be limited.

Supplementary analyses

Amino acids

The determination of amino acid (AA) concentrations in feed proteins first requires hydrochloric acid hydrolysis of the proteins to release the individual AAs. Depending on the sample being analyzed, among other factors, the hydrolysis procedure must be modified for the accurate determination of specific AA. The hydrolysate is then subjected to ion exchange or high performance liquid chromatography (HPLC) coupled with spectrophotometric or fluorometric detection. The complexity of the methods and the cost of the equipment required means the analysis is relatively expensive and seldom justified for routine use. Reference values for common feeds are listed in Appendix A, Tables A2.4.

Figure 3.8: An automated HPLC amino acid analyzer.

Fatty acids

The fatty acid (FA) profile of feeds is determine by gas chromatography after releasing the individual FA from triglycerides and other complexes by saponification—treatment with sodium hydroxide as used in making soap. The free fatty acids are then converted to methyl esters before being subjected to chromatography. FA profiles of feeds commonly used in western Canadian dairy diets are listed in Appendix A Table A2.5.

Vitamins

Vitamins are small organic molecules and, as such, they are usually measured by gas liquid chromatography (GLC) or high performance liquid chromatography (HPLC). Because these procedures are relatively expensive, vitamin levels in feed are not routinely measured.

Values calculated from analysis results

Non-fibre carbohydrates (NFC; line 14)

Non-fibre carbohydrates (NFC) represent feed carbohydrates, including starch, pectin and sugars, which are more rapidly degradable by digestive tract microbes compared to the cell wall carbohydrates measured as neutral detergent fibre (NDF).

NFC concentration is calculated by difference:
$$NFC\% = 100\% - CP\% - EE\% - NDF\% - Ash\%$$

A variation of NFC is non-structural carbohydrates (NSC; line 14), where:
$$NSC\% = 100\% - CP\% - EE\% - (NDF\% - NDF\text{-}P\%) - Ash\%$$

Adjusted crude protein (ACP; line 15)

Adjusted crude protein (ACP) discounts CP to account for ADF-N. In most feeds, 3-8% of total CP will be associated with the ADF residue, even in the complete absence of heating. Therefore, most feed labs do not discount the total CP value for heat damage unless ADF-N values are excessive. Others assume that a fixed proportion (e.g., 70%) of ADF-N is unavailable.

Discounted CP (Total CP – ADF-P or Total CP – excess ADF-P) values are generally reported as adjusted crude protein (ACP) or available protein (AP).

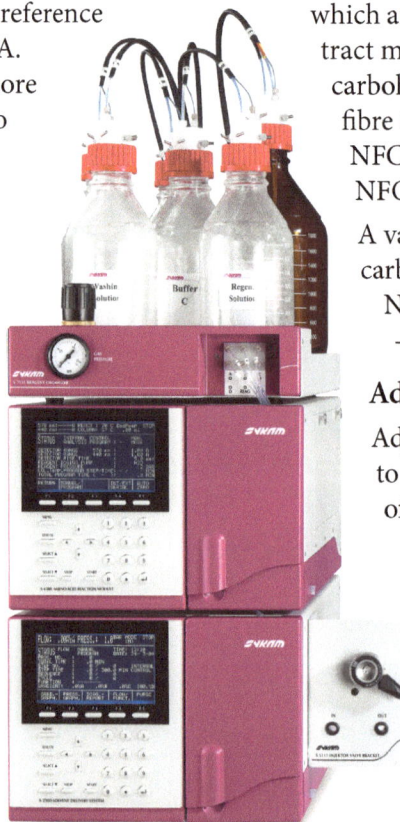

Relative feed value (RFV; line 15)

Relative feed value is used as an index of forage digestibility and potential intake, calculated from ADF and NDF using the following equation:

$$RFV\% = \frac{[(88.9 - (0.78 \times ADF\%)) \times (120 / NDF\%)]}{1.29}$$

Digestible protein (DP; line 16)

Digestible protein is typically calculated from CP assuming a true digestibility of 90 percent and a correction of 3 percent for metabolic fecal protein:
$$DP\% = (CP\% \times 0.9) - 3\%$$

Alternative calculations to estimate DP, based on recommendations for beef cattle may also be used. For example:
$$DP\% = 72.96\% - (1.02 \times ADF\text{-}P\% \times 100\%/CP\%)$$

Metabolizable protein (MP; line 16)

Metabolizable protein is usually calculated as 70 percent of DP:

$$MP\% = (DP\% \times 0.7)$$

Total digestible nutrients (TDN; line 18)

The classical method of estimating TDN was based on the summation of digestible fractions determined by proximate analysis (page 33):

$$TDN\% = \text{dig CP\%} + \text{dig CF\%} + \text{dig NFE\%} + (2.25 \times \text{dig EE\%})$$

The digestible ether extract is multiplied by 2.25 because, on oxidation, lipids provide approximately 2.25 times more energy per unit than the other fractions.

This method was very laborious and impractical for routine use, requiring the estimation of digestibility in animal feeding trials where each fraction was measured in feeds and feces over several days. Today, feed TDN value is usually calculated from ADF as described below for other energy fractions.

Gross, digestible, metabolizable, net energy (lines 18-20)

The relationships between various energy fractions are described in Chapter 5 (pages 56 - 57). Gross energy (GE) potential of a feed is estimated by measuring the heat produced by complete combustion of a sample under controlled conditions in a 'bomb' calorimeter.

Digestible energy (DE) was originally estimated as the gross energy measured in the total amount of feed consumed over several days minus the gross energy measured in the total feces produced in the same time period, similar to the methods described for the determination of TDN.

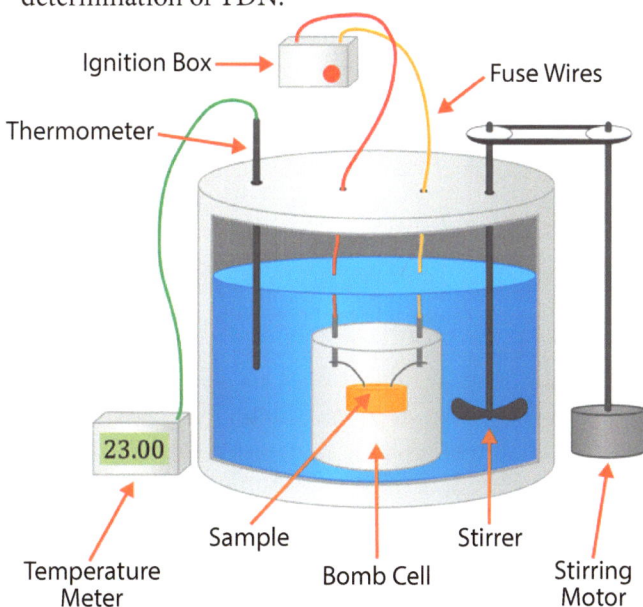

Ignition Box

Fuse Wires

Thermometer

23.00

Temperature Meter

Sample

Bomb Cell

Stirrer

Stirring Motor

Figure 3.9: Schematic diagram of a bomb calorimeter.

Estimates of the digestible (DE), metabolizable (ME) and net (NE) energy values of forages are most commonly based on measurements of their acid detergent fibre (ADF) content. Since ADF represents the most indigestible fraction of a feed, it is assumed that the digestibility of a feed, and thus its energy-yielding potential, should be inversely proportional to its ADF content (figure 3.10 page 38).

Here are examples of equations used to calculate DE:

for legume forage:

$$DE, Mcal/kg = 3.91 - 0.036 \times ADF\%$$

for grass forage:

$$DE, Mcal/kg = 4.34 - 0.046 \times ADF\%$$

for mixed grass/legume forage:

$$DE, Mcal/kg = 4.08 - 0.040 \times ADF\%$$

Equations used vary from lab to lab, so it is important to determine whether differences in reported energy values between labs is due to different ADF values or different equations or both.

Estimates of the energy contributed by concentrate feeds are most commonly based on values published in tables of feed composition (see Appendix A).

The following equation is recommended in NASEM Dairy 8 to convert DE to ME:

$$ME, Mcal/kg = 0.82 \times DE, Mcal/kg;$$

The conversion of ME to NE values varies with the productive or reproductive state of the animal as well as the energy density and nutrient composition of the diet. For growing and lactating dairy cattle the following simple conversion can be used:

$$NE, Mcal/kg = 0.66 \times ME, Mcal/kg.$$

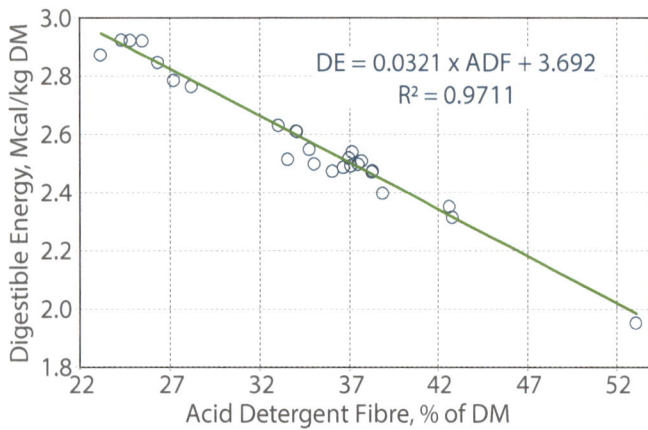

Figure 3.10: *The relationship between Acid Detergent Fibre (ADF) and Digestible Energy (DE) for all grain crop forages in Appendix table A1.2. In this example, ADF accounts for 97% of the variation ($R^2 = 0.97$) in DE.*

Digestion analyses

In situ protein degradability

Protein digestion in ruminant animals is a two-stage process (see figure 4.14, page 50). First, a fraction of the dietary protein is broken down by microbes in the rumen to produce peptides, amino acids and ammonia. This fraction is referred to as rumen degradable protein (RDP). The dietary protein that escapes breakdown by rumen microbes (commonly called 'bypass' protein) is referred to as rumen undegradable protein (RUP).

The potential degradability of protein in the rumen is estimated by measuring the disappearance of crude protein (CP) from finely ground feed samples incubated in porous nylon bags in the rumen of a

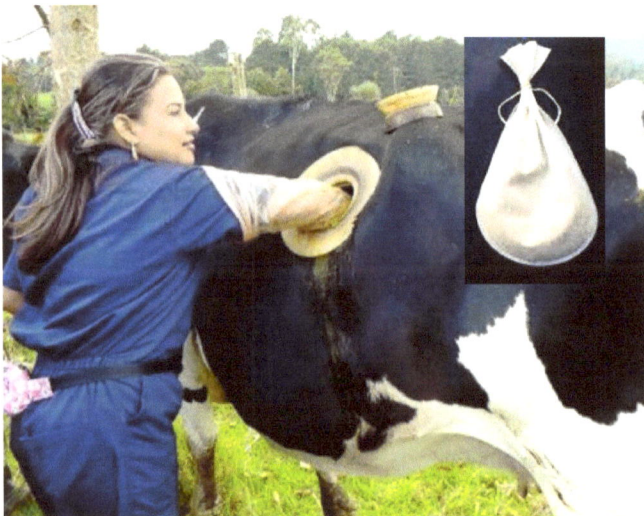

Figure 3.11: *The nylon bag method, used for estimating ruminal protein degradability, requires the rumen incubation and sequential removal of multiple finely ground feed samples contained in porous nylon bags.*

fistulated animal (figure 3.11). CP remaining in bags removed from the rumen at fixed time intervals allows estimation of the A, B and C fractions as well as the rate of degradation of the B fraction (k_d of B, %/hr) shown in the table of feed analyses (Appendix A, table A1.1). As illustrated in figure 3.12, the A fraction is the CP that washes out of bags before microbial degradation begins; the C fraction is the CP that remains in the bag regardless of the length of incubation; the B fraction is total CP minus the sum of fractions A and C. The rumen undegradable protein (RUP) values in Appendix A, tableA1.1 are calculated from the results of in situ degradability trials combined with assumptions about passage rates of specific feeds through the rumen.

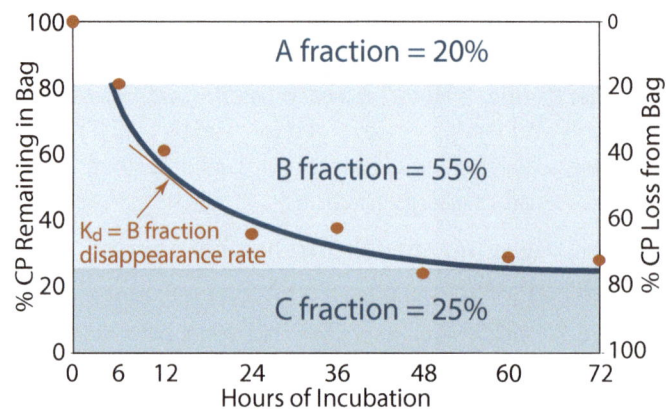

Figure 3.12: *Data points reflecting disappearance of protein from nylon bags incubated in the rumen.*

The in situ procedure is laborious and expensive, limiting its use to research facilities. Results obtained have been quite variable both within and between laboratories. Simpler benchtop methods have been proposed, in which feed samples are incubated with mixtures of protein-degrading enzymes extracted from the rumen. Although several commercial labs offer degradability analysis using these methods, lack of standardization of both protocols and enzymes makes it difficult to place confidence in results.

In vitro digestibility

The rumen digestibility of neutral detergent fibre (NDF) or dry matter (DM) is estimated by incubating a finely ground feed sample with a rumen inoculum in a flask for 30 or 48 hours under anaerobic (no oxygen) conditions. In vitro NDF digestibility (IVdNDF%) is calculated as the difference between the total NDF of the original sample (NDFtotal) and that of the residue recovered through filtration of the flask contents after incubation (NDFresidual):

$$IVdNDF\% = \frac{NDFtotal - NDFresidual}{NDFtotal} \times 100$$

In vitro dry matter digestibility (IVdDM) is estimated by the same method, simply substituting DM for NDF in the above equation:

$$IVdDM\% = \frac{DMtotal - DMresidual}{DMtotal} \times 100$$

Near infrared reflectance spectroscopy (NIRS)

NIRS is a quick, reliable, low cost, computerized method used to analyze feeds for their nutrient content. An NIR spectrometer (figure 3.13) uses reflected light in the near infrared part of the spectrum instead of chemicals to identify important compounds and measure their amounts in a sample. Feeds can be analyzed in less than 15 minutes where chemical methods may take hours or days. This quick turnaround and the resultant cost savings in labour and consumables make NIRS an attractive method of analysis.

Figure 3.13: A near infrared reflectance spectrometer.

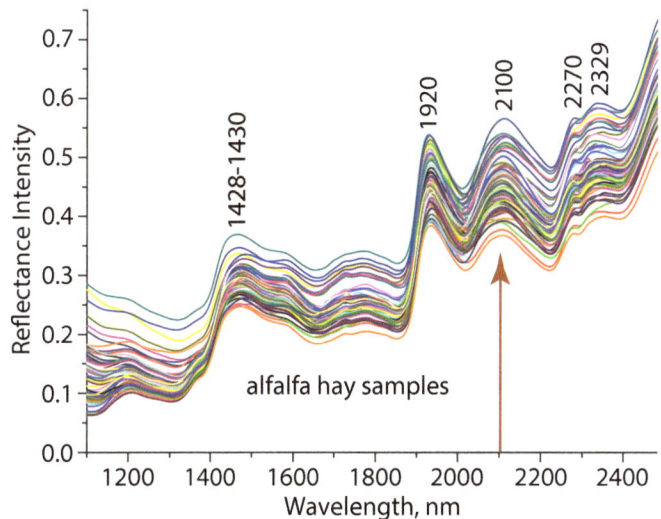

Figure 3.14: Near infrared reflectance spectra for a series of alfalfa hay samples.

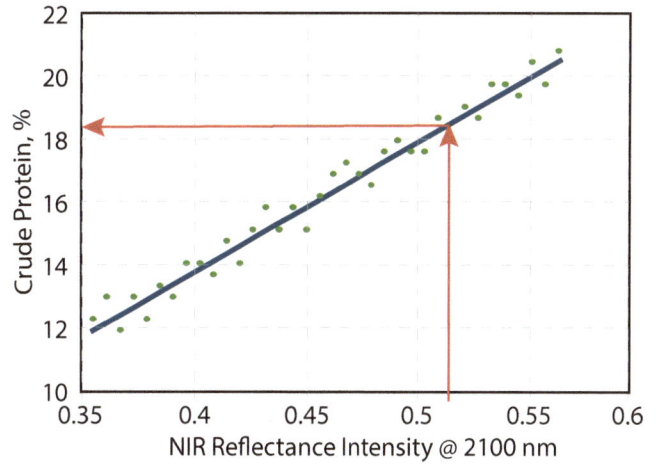

Figure 3.15: A calibration curve relating crude protein concentrations of alfalfa hay samples measured in a feed lab with the near infrared reflectance intensity at 2100 nm of the same samples shown in figure 3.14.

The use of NIRS requires the establishment of calibration equations for each type of feed being analyzed, relating specific reflectance wavelengths with chemical analysis results for the same feed type. Figure 3.14 illustrates near infrared reflectance spectra for a series of alfalfa hay samples. A suitable calibration equation might relate reflectance at 2100 nm with feed crude protein concentration (see fig 3.15).

On-farm feed quality assessment

Dry matter and moisture

The laboratory assessment of feed dry matter and moisture content was described earlier (page 32). When feeding ingredients such as silage or wet brewer's grains, it is often necessary to quickly check moisture levels to make sure the correct amount of ingredient dry matter is being included in the ration. The sidebar on the next page describes how to check moisture levels with a microwave oven. Another alternative is the Koster Moisture Tester, illustrated in figure 3.16.

Silage pH

The stable preservation of silage is dependent upon its moisture content and acidity. The latter is measured on the scale of pH. Pure water has a pH of 7. A pH of 6 indicates a low level of acidity while pH 5 is ten times as acidic as pH 6 and pH 4 is again ten times as acidic as pH 5. pH values above 7 indicate alkalinity.

After fermentation is complete, high moisture silage (60 – 75% moisture; 25 – 40% DM) should have a pH below 4.5. The pH of haylage or low moisture silage may be slightly higher. Of particular concern to dairy farmers, poorly preserved silage may cause outbreaks

Microwave oven dry matter estimation

Forage dry matter levels can be accurately estimated on-farm using an inexpensive microwave oven and an electronic postal scale. Mechanical postal scales are generally not accurate enough to measure gram differences in weights.

Here's how it's done :

1. Weigh a microwave-safe container large enough to hold 100 – 200 grams of wet forage (a paper bag is a good choice). Record the weight of the container (WC) or, if your scale has a tare adjustment, set the scale at zero (WC = 0).
2. Weigh 100 – 200 grams of wet forage into the container (WW). The larger the sample, the more accurate your determination can be.
3. Place a drinking glass or glass jar containing 250 ml of water in the back corner of the oven. The water serves as a 'ballast' to absorb excess energy, preventing ignition of the sample. If your sample does ignite, turn off the oven, unplug the power but don't open the door until the sample has burned completely.
4. Heat the forage sample at 80 – 90% of maximum power for 5 minutes. Re-weigh and record the weight.
5. Repeat step 4 until the weight is less than 5 grams lower than the previous weight.
6. Heat the sample at 30 – 40% of maximum power for 1 minute. Re-weigh and record the weight.
7. Repeat step 6 until the weight is less than 1 gram lower than the previous weight. This is the dry weight (WD).
8. Calculate Dry Matter (DM) % as follows:

$$DM\% = \frac{WD - WC}{WW - WC} \times 100$$

Figure 3.16: The Koster Moisture Tester. A weighed sample of feed placed in the basket is dried to constant weight by placing the basket and sample on the fan-driven heater.

Bulk density (bushel weight)

Bulk density is the weight of a standard volume of grain. In the past, this was measured as bushel weight but, since adoption of the metric system, kilograms per hectolitre (kg/hL) is used as the standard unit of measure. Table 3.1 provides bulk density standards for western Canadian feed grains. The Avery bushel used in Canada is equivalent to 36.37 litres. Be aware that, in the US, the 35.34 litre Winchester bushel is used.

Although bulk density is a generally accepted measure of grain quality, its relationship to nutritional value is not clear. Most assume that grain of higher density is worth more as a dietary ingredient due to the fact that the grain kernel is denser than the hull and also higher in digestible nutrients (mainly starch). The practical implication of this relationship is very useful when a producer is buying grain without the benefit of a feed test. Light grain (e.g., barley weighing 42 lbs/bushel) should be worth less than heavy grain (e.g., barley weighing 50 lbs/bushel).

	lbs per bushel	kgs per hectolitre
Barley	48	62
Oats	37	48
Rye	49	63
Triticale	48	62
Wheat	51	65

Table 3.1: Bulk density standards for Canadian feed grains.

of listeriosis or circling disease (see page 21). When high moisture forage is not well chopped prior to ensiling, it is impossible to eliminate sufficient air when packing. Inadequate fermentation results in a pH which is too high, providing good conditions for the proliferation of the bacteria that cause the disease (*Listeria monocytogenes*).

Silage pH can be monitored with a simple, portable pH meter or with pH test strips, preferably strips designed to measure a pH range between 4.0 and 7.0 with distinct colour differences in that range.

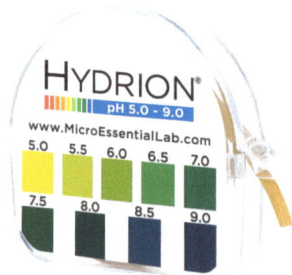

Figure 3.17. pH test strips

Figure 3.17: The Penn State Particle Separator.

Particle size

Cattle diets must include structured (physically effective) fibre to promote chewing, rumination, saliva production and particle stratification in the rumen (see pages 43 - 44). The Penn State particle separator provides estimates of particle size distribution as an aid to evaluating dietary adequacy of physically effective fibre.

The separator, illustrated in figure 3.17, consists of four trays with the top three having 19 mm (0.75 in), 8 mm (0.31 in) and 1.18 mm (0.046 in) holes, in order from the top to the bottom tray. The bottom tray has a solid bottom.

After placing approximately 1.4 L of forage or total mixed ration on the top screen, the trays are shaken vigorously on a flat surface five times in one direction, then turned one-quarter turn, and shaken 5 times again. This process is repeated eight times to give a total of 40 shakes after which the material on each tray is weighed. The percentage of the total sample dry matter on each tray can be determined using a microwave oven or Koster moisture tester, as described on page 40.

Sugar content of crops

One of the key factors in successful silage making is an adequate water-soluble carbohydrate (WSC) concentration in the crop when it is ensiled. Bacteria ferment the WSC to produce lactic acid, causing the pH of the silage to decrease. The higher the WSC content of the crop, the more rapid the decline in pH once the silage pack has reached anaerobic conditions.

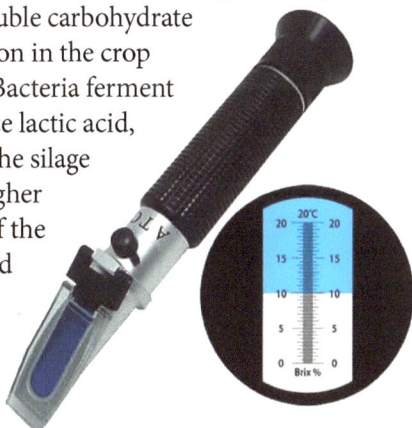

Figure 3.18: A Brix refractometer

Crop WSC content is normally highest in the afternoon of a sunny day, due to the rapid rate of photosynthesis resulting from increasing light intensity. To evaluate the crop WSC concentration, a Brix refractometer can be used. Liquid can be extracted from a small sample of the standing crop using a household garlic press. A drop of the liquid extract placed on the refractometer lens will produce a Brix reading as shown in figure 3.18. Brix readings are relative; experience will reveal the normal range for specific crops and growing conditions. For example, a Brix level of 22 is high for alfalfa, while 8 is average; for silage stage cereal crops, 18 is high and 10 is average.

Factors affecting feed quality

Several factors that affect feeding value have already been mentioned:
- the energy content of feed grain is roughly proportional to bulk density (bushel weight);
- heating in storage reduces protein availability in hay and silage;
- improper preservation of silage results in lower energy and protein values.

In addition to these, stage of maturity, legume content and weathering can have major effects on forage quality.

Stage of Maturity at Harvest	ME[†] Mcal/kg	% Crude Protein grass	% Crude Protein legume	Intake % BW
Vegetative	2.3	15	21	3.0
Boot or Bud	2.1	11	16	2.5
Bloom	1.8	7	11	2.0
Mature	1.6	4	7	1.5

Table 3.2: Effect of stage of maturity on the feeding value of grass and legume forages. [†]ME - metabolizable energy.

The optimum time for harvest is a compromise between quality and quantity as shown in figure 2.1 on page 23. The target will be based primarily on the nutrient requirements of the animals to which the forage will be fed. For example, a dry cow might require a forage containing less than 12% crude protein, whereas the same cow in early lactation will be fed forage with a much higher protein concentration.

Table 3.2 illustrates differences in protein content of grasses and legumes at various stages of growth. Forages which have significant legume content have higher feeding values than those containing grasses alone. Finally, nutrients can be lost from hay during any stage of harvest or storage (table 2.1, page 24):
- cellular respiration: 4 - 15% of initial dry matter may be lost during wilting;

- legumes are particularly susceptible to leaf shattering if the crop has to be raked or tedded, and during baling, especially if the crop is very dry. Since leaves are significantly higher than stems in nutrient content at all stages of maturity (see figure 1.8, page 8), it is not surprising that leaf loss leads to lower forage quality.
- rain damage can leach nutrients from the crop as it lies in the field or in storage if it is not covered;
- vitamins, particularly vitamin A, may be lost due to UV exposure in the field;
- if moisture content exceeds 15 – 20%, heat produced by cellular respiration and/or microbial activity can result in the formation of heat-damaged protein (see ADF-N in Feed Analysis section, page 33), mould growth and spontaneous combustion.

Chapter 4: The Digestion of Feed

The digestive systems of mammals are broadly divided into three classes:
- monogastric – meaning one stomach, includes: human, dog, cat, pig;
- ruminant – includes: cattle, sheep, goats, deer, bison;
- hindgut fermenters – includes: horse, rabbit.

In order to understand some of the unique properties of the ruminant system, it will be helpful to first briefly describe the simpler monogastric system.

Monogastric digestion

An outline of the pig's digestive system is shown in figure 4.1. Food, upon entering the mouth, is pulverized by chewing. At the same time, lubricating digestive juices containing enzymes are secreted from the salivary glands. These particular enzymes are responsible for initiating the breakdown of starch.

Food, now mixed with these secretions, passes down the esophagus into the stomach, where the digestion of carbohydrates, proteins, lipids and other organic compounds is initiated by acid and other specific enzymes. In monogastrics, the stomach also serves as a reservoir for food that has been rapidly ingested.

When the initial stages of digestion in the stomach are completed, the contents pass into the small intestine. Here, bile from the liver and gallbladder, as well as enzymes from the pancreas are added. Breakdown continues as the digesta travel the length of the small intestine, while at the same time the products of digestion are absorbed into the bloodstream. Enzymatic breakdown of most of the organic constituents of food is

Compartment	Digestive Volume (% of total tract)			
	Cattle	Sheep	Horse	Pig
Rumen	56.9	52.9		
Reticulum	2. 1	4.3		
Omasum	5.3	1.7		
Abomasum	6.5	7.7		
Total Stomach	70.8	66.6	8.6	29.2
Small Intestine	18.5	20.5	30.2	33.3
Cecum	2.8	2.6	15.9	5.6
Large Intestine	7.9	10.3	45.3	31.9
Total Capacity (litres)	356.0	44.0	211 .0	28.0

Table 4.1: Comparison of the relative sizes of digestive tract compartments in adult animals.

complete by the time the unabsorbed digesta reach the cecum and large intestine. One of the main functions of this part of the system is the absorption of water and minerals of both dietary and secretory origin. Further breakdown of digesta is carried out here by a permanent population of microbes (bacteria, fungi and protozoa) with some proportion of the products being absorbed into the blood. Food material which has escaped both enzymatic and microbial digestion is excreted.

Hindgut fermenters, like the horse and rabbit, have a relatively large capacity for microbial digestion in the large intestine and a much-enlarged cecum. Table 4.1 compares the sizes of several parts of various digestive systems. Notice that where the large intestine and cecum make up only 10.7% of the digestive system in cattle, they represent over 60% of the total volume in the horse. The implications of this will become clear when we discuss microbial digestion in the ruminant.

The ruminant digestive system

The main feature that distinguishes the monogastric digestive system from that of dairy cattle is the complex stomach illustrated schematically in figure 4.2. The first two sections, the reticulum and rumen, comprise a large fermentation compartment, referred to as the reticulorumen. These two sections are distinguishable by their unique linings—the inner surface of the reticulum has a 'honeycomb' appearance whereas the lining of the rumen is like pile carpet having innumerable small, flat projections called papillae (see figure 4.3).

Ingested feed passes rapidly, with very little chewing, down a muscular esophagus into the reticulorumen. Later, boluses (cuds) of feed are regurgitated, broken down by re-chewing (rumination) and mixed with saliva.

Ruminant animals require 'physically effective fibre' in their diets to keep the rumen functioning normally. In addition to breaking down large forage fibre particles,

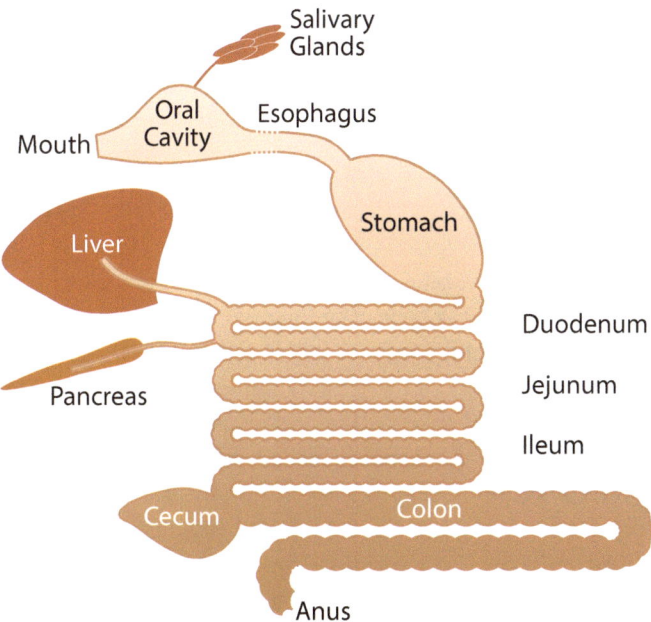

Figure 4.1: Schematic of the monogastric digestive system.

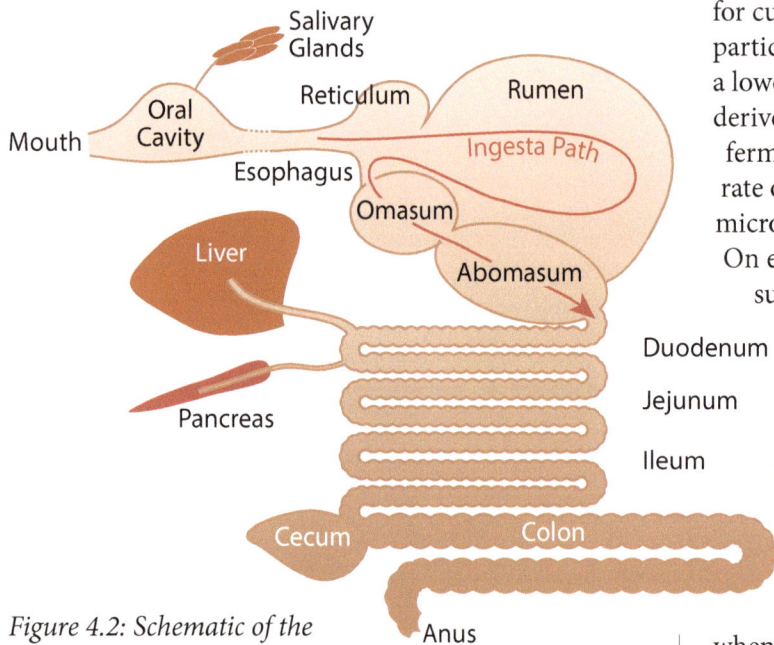

Figure 4.2: Schematic of the ruminant digestive system.

chewing and rumination promote salivation. A mature cow produces up to 200 litres of saliva per day which has three main functions:

- it lubricates feed as it moves down the esophagus;
- it provides the liquid flow which carries feed particles into the omasum, and;
- it contains digestive enzymes (e.g., amylases) as well as buffers which prevent rumen pH from falling too low as microbes produce acids from feed fermentation.

Figure 4.3: Inner surfaces (mucosa) of the reticulum (left) and rumen (right).

Physically effective fibre also provides a 'tickle factor' which stimulates rumen contractions. These help to keep the rumen contents well mixed and to force fluid and small particles further down the digestive tract. A third function of long forage is the maintenance of a fibre mat (figure 4.4) which floats in the rumen and functions as a particle sorting system. Long particles near the top of the mat are the first to be regurgitated

for cud chewing which subdivides and adds saliva to the particles. When they re-enter the rumen, they 'float' at a lower level than the longer, drier particles they were derived from. A functional mat also stabilizes rumen fermentation by trapping fine particles, slowing their rate of breakdown by reducing exposure to rumen microbes.

On entering the reticulorumen, feed is immediately subjected to microbial digestion. An extremely varied population of bacteria, protozoa and fungi (figure 4.5) attach themselves to the feed and begin the breakdown process. This is facilitated by the secretion of enzymes onto the feed and into the fluid contents of the rumen.

It should be noted that the microbial population that attaches to a particle of grain will be quite different from that which attaches to a forage leaf. This has important implications when changes in ration are contemplated. A slow transition is necessary to allow time for microbial populations to adapt to new feeds.

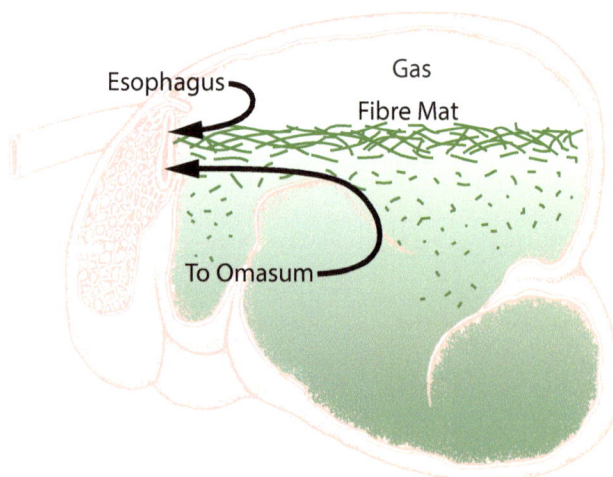

Figure 4.4: The fibre mat is important for the maintenance of rumen function and efficient fermentation.

The convoluted inner surfaces of the reticulum and rumen serve two main functions. They vastly increase the area for absorption of nutrients and they also provide attachment sites for additional populations of bacteria. These bacteria, like the ones attached to feed particles, produce enzymes which are secreted into the fluid contents of the rumen. One of the important contributions of this particular population is the enzyme urease which is responsible for the breakdown of urea. So feed is subjected to digestion both by enzymes secreted into the general milieu of the rumen and, more specifically, by those produced by attached microbes.

Continual mixing of rumen contents is essential to efficient fermentation. The muscular walls of the rumen and reticulum produce waves of contraction traveling

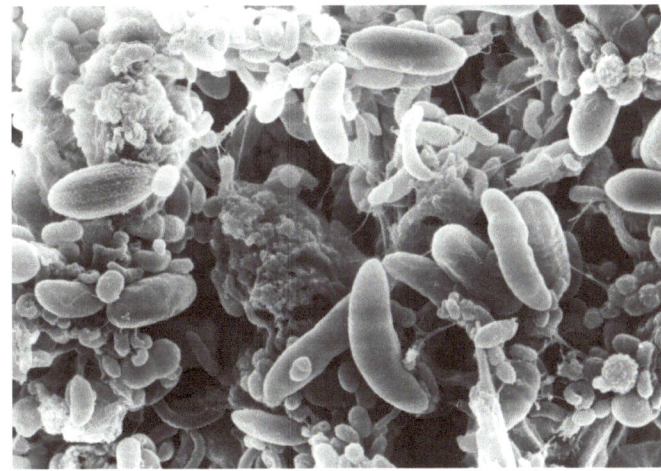

Figure 4.5: Scanning electron micrograph showing the variety of microorganisms found in the rumen.

their combined lengths at about half-minute intervals. This process, in addition to mixing the rumen contents, facilitates both regurgitation for further 'cud-chewing' and belching, which releases gases produced by fermentation (mainly hydrogen and methane). Under some conditions (e.g., grain overload) the muscular walls may stop contracting resulting in rumen stasis, which can place the animal at serious risk of bloating.

After the feed has been sufficiently chewed and broken down by microbial action, the digesta enters the omasum. Flow into this third segment of the ruminant stomach is regulated by a small opening called the reticulo-omasal orifice which prevents large particles from leaving the rumen. It is the small caliber of this orifice which makes it possible for small ruminants (e.g., sheep) to utilize whole grains. The larger orifice in cattle allows particles the size of whole grain to pass into the lower gut and be excreted.

The omasum itself is a muscular organ having multiple internal leaves (laminae) like the pages in a book (see figure 4.6), providing a large internal surface area. It is thought to have two main functions. The first is

Figure 4.6: The 'many-leaved' lining of the omasum (left) and the mucosal lining of the abomasum (right).

the absorption of water, electrolytes and products of microbial digestion from the rumen fluid, yielding a product for further digestion which has a significantly higher proportion of dry matter. Secondly, the omasum serves as a pump, propelling digesta from the reticulorumen into the fourth segment of the stomach, the abomasum.

The ruminant abomasum is analogous to the true stomach of the monogastric with its digestive processes being very similar to those described earlier for the pig. Digestion and absorption of its products progress as the digesta passes into and down the small intestine.

The large intestine and cecum of cattle represent only about 11% of the total volume of its digestive system. This may seem relatively insignificant in comparison with the horse (table 4.1). However, fermentation in this area can make a significant contribution to overall digestion. This will be discussed further in the section on carbohydrate digestion (page 47).

Development of the ruminant stomach

At birth, the calf's rumen and reticulum have a capacity roughly equal to that of the abomasum (figure 4.7). They contain no micro-organisms and, as a consequence, are not capable of functioning as they do in the adult. Bacteria begin to populate the rumen shortly after birth as the calf begins to feed and sample its environment. However, it takes several weeks before a stable microbial population is established which is capable of efficient digestion.

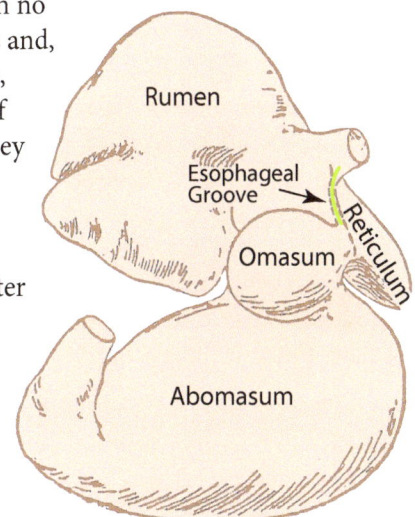

Figure 4.7: The stomach of the newborn calf.

Attempts have been made to hasten the establishment of functional microbial populations in newborn calves by inoculation with mixed microbes from the rumens of mature animals. There are two reasons for this. First, many of the bacteria which contaminate the digestive tract from the environment early in life are capable of producing digestive upsets. By inoculating the rumen and reticulum with a more appropriate microbial population, competition might protect the digestive tract from the adverse effects of such contaminants.

The second reason for attempting to establish a functional population is to hasten the ability of the rumen and reticulum to digest solid feed. This would make it possible to wean calves earlier, a particular advantage when the cost of liquid feeds (e.g., milk replacer) is much higher than that of solid feeds.

The esophageal groove

Since the rumen and reticulum are non-functional in the newborn calf, a mechanism has evolved which allows milk to flow directly to the omasum. A reflex reaction causes a muscular fold on the wall of the reticulum to form a closed tube leading from the end of the esophagus to the reticulo-omasal orifice (see figure 4.7). This fold is called the esophageal groove and an appreciation of its function will affect some of the management aspects of feeding newborn calves.

The esophageal groove closes in response to behavioural stimuli associated with the ingestion of liquid feed such as nursing from the cow or feeding from a nipple pail or bottle. Even the sight of a nipple bottle may elicit the response. However, the reflex requires some degree of training. Therefore, it is used to best advantage with regular feeding routines.

If at all possible, weak newborn calves should be encouraged to suckle either from the cow or from a bottle. Although feeding by stomach tube may be the only alternative in some cases, this will often result in milk spilling into the reticulorumen. A similar situation arises when milk is ingested too rapidly to be accommodated by the esophageal groove. This can occur when milk replacer is fed from a bottle or from the bottom of a nipple pail where a round-holed (rather than a crosscut) nipple is used.

Milk that finds its way into the rumen and reticulum is subjected to fermentation by bacterial contaminants early in life. Such fermentation may result in significant gas production resulting in a typical pot-bellied calf. The young calf cannot expel this gas efficiently since the belching mechanism is poorly developed.

Effect of feeding management

Between birth and maturity, the rumen and reticulum increase tenfold in volume in relation to the abomasum; the rate at which this proceeds can be significantly altered by nutritional management.

Most newborn calves show little interest in consuming solid feed before they are two or three weeks of age. Consequently, until that time they must be nourished exclusively with milk or milk replacer. After this time it is possible to accelerate rumen development through feeding practices.

The closure of the esophageal groove only occurs when liquid feed is ingested. Therefore when solid feed is consumed it travels directly to the rumen where it is fermented to produce volatile fatty acids (VFA; see section on Carbohydrate digestion, page 47).

In the past, it was commonly believed that feeding hay to young calves would promote rumen development. The rationale was that physical 'scratch' was needed to start the rumen working. It is now known that the main stimulus to rumen development is VFA production from feed fermentation.

Because the amounts of VFA produced from grain are higher than those from forage, the rumen develops much faster when grain-based diets are fed, as shown in figure 4.8. For this reason, hay should not be offered to calves until after they are weaned. By the time the calf has become a true ruminant, physically effective fibre is required to promote the growth of the muscular layer of the rumen and to maintain the health of the papillae. On all-concentrate diets, rumen papillae can grow too rapidly in response to high levels of VFA. When this happens, they may clump together, reducing the surface area available for absorption.

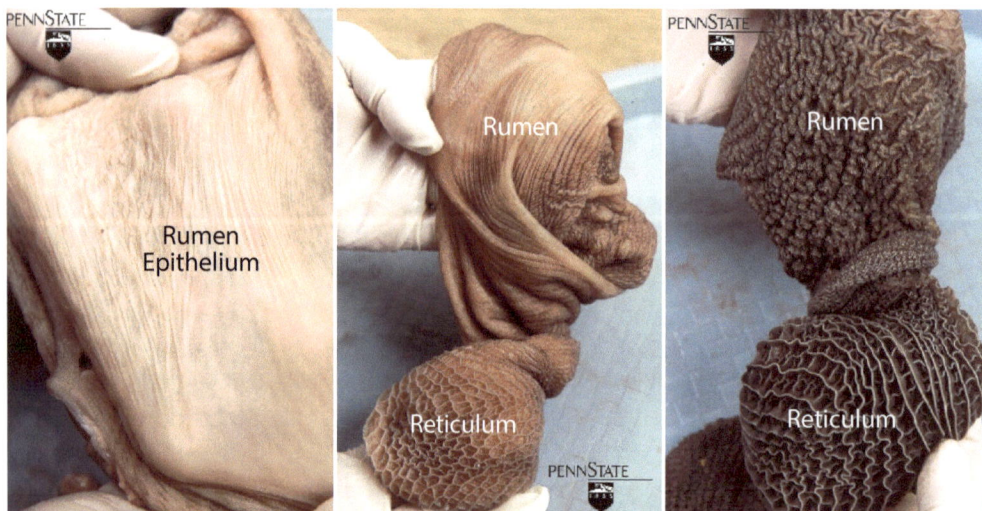

Figure 4.8: Differences in development of the rumen in six week old calves fed milk only (left), milk and hay (centre) or milk and grain (right). source: Penn State University

Also, some 'scratch' is needed to keep the papillae from developing an outer layer of keratin (skin-like tissue), which can also inhibit VFA absorption.

Carbohydrate digestion

As pointed out earlier, forages contain only low levels of fats and oils (lipids). Consequently, the main sources of energy for the ruminant are the carbohydrates.

Monogastric carbohydrate digestion

Carbohydrate digestion in the monogastric begins when food is mixed with saliva containing enzymes (amylases) which begin the breakdown of starch. The process continues in the small intestine, facilitated by pancreatic amylases. As digestion progresses, the end products (the simple sugars: glucose, galactose, etc.) are absorbed into the bloodstream. Depending on the energy status of the animal, the sugars may be used as immediate energy sources or stored as glycogen for later use.

Ruminant carbohydrate digestion

The enzymes that mediate carbohydrate breakdown in the ruminant are mainly of microbial origin. Each type of carbohydrate is digested by specific enzymes produced by a distinct microbial population.

Volatile fatty acids

If oxygen were present in the rumen, the end-products of carbohydrate digestion would be carbon dioxide and water, the compounds from which the carbohydrates were originally synthesized in the plant (page 3). However, the microbial population in the rumen operates in the absence of oxygen (the rumen environment is *anaerobic*), resulting in incomplete carbohydrate breakdown. Under these conditions, the principal end-products of digestion are volatile fatty acids (VFAs), including acetic acid, propionic acid and butyric acid. Incomplete breakdown in this case is analogous to the situation where wood is burned with limited air. The smoke produced is composed of the products of incomplete combustion.

The breakdown of carbohydrates to VFAs results in the release of significant amounts of feed energy. This energy is utilized in the rumen for microbial growth involving, for example, the synthesis of new microbial protein, fat and carbohydrate. The VFAs are absorbed into the bloodstream through the wall of the rumen to serve as energy sources for the animal itself.

Other important products of ruminant carbohydrate digestion are the keto acids. These are formed in much smaller quantities than the VFAs, but serve an important role in microbial protein synthesis (page 50). Methane, also a major end-product of anaerobic fermentation, is largely lost to the atmosphere.

Carbohydrates that escape digestion in the rumen and those associated with the microbial population

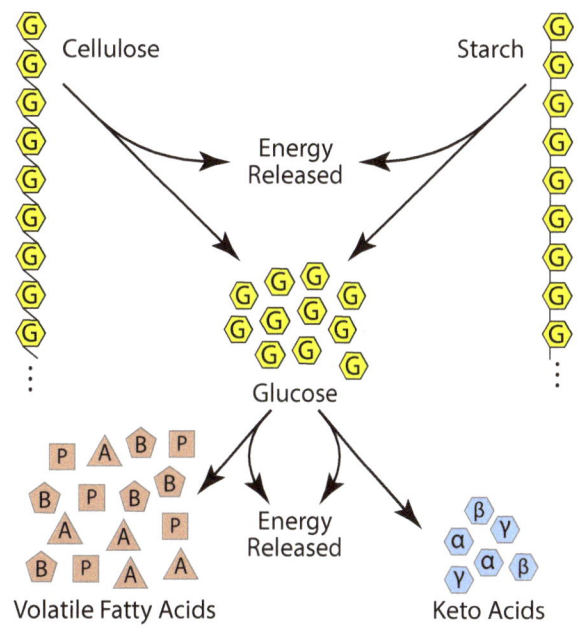

Figure 4.9: *In the ruminant, microbes initially break down both fibrous and non-fibrous carbohydrates to produce simple sugars which are subsequently fermented to produce VFAs, keto acids and other end-products.*

may be utilized further down the digestive tract. For example, carbohydrates that are components of microbial cells are digested and the resulting sugars absorbed in the small intestine, as in monogastrics. The contributions of these simple sugars to the overall energy requirements of the ruminant are minor. In the lactating cow, absorbed glucose is primarily used in the manufacture of the milk sugar, lactose.

Further carbohydrate breakdown also occurs in the large intestine. Experimental results suggest that 5–10% of the energy requirement of calves could be met from VFAs produced by the microbial population found here.

Cellulose digestion

The microbial populations of both the rumen and the large intestine produce the enzyme cellulase which is responsible for the breakdown of cellulose. It is this feature which makes ruminants and hindgut fermenters like the horse unique in their ability to utilize forages in the production of meat, milk and fibre. Mammals are incapable of digesting cellulose without the aid of these microbes. Two features of cellulose utilization should be appreciated when feeding dairy cattle:

The digestion of cellulose is a relatively slow process. One implication of this is the fact that feed consumption is limited by the rate at which feed is digested. Feed can be introduced into the rumen only as rapidly as the products of digestion are discharged. This means that rations high in fibre will be consumed in lower quantities than those that are low in fibre.

A second consideration is the degree to which the cellulose in a feed is associated with lignin (see figure 1.3). As forages mature, the cellulose found in the plant's cell walls becomes more lignified. The result is that the cellulose becomes less digestible. This is reflected in lower energy values and intake potential for forages as they mature (table 1.1, page 4).

Other carbohydrates

Starch, pentosans and simple sugars are rapidly fermented in the rumen to produce VFAs, keto and lactic acids, CO_2, hydrogen and methane. Under most conditions, lactic acid production is low. However, when feeds containing large quantities of readily fermentable carbohydrates are rapidly consumed, lactic acid production may be significantly higher, resulting in ruminal acidosis (pH below 5.5). Since stable pH is critical to proper rumen function, rapid changes can cause digestive problems. For example, acute grain overload results in depressed rumen pH which can lead to rumen stasis. The animal becomes unable to expel fermentation gases by belching and bloat occurs. The effects of prolonged excessive grain intake are also seen in the health of the rumen papillae which can become damaged (figure 4.10) with long-term exposure to low rumen pH and/or inadequate stimulation from physically effective fibre.

Figure 4.10: Healthy (left) and acidosis-damaged (right) rumen papillae.

The effect of physical structure

Carbohydrate degradation rates in the rumen also depend upon the physical structure of feeds. For example, unprocessed grain is poorly digested in the rumen because the starch contained in the grain kernel is inaccessible to microbial attack. Conversely, the starch in finely ground grain is degraded very rapidly (see Processing Index, page 28). Although the differences are not as great, fine chopping and aggressive crushing of forages will also increase rates of degradation.

Non-structural carbohydrates that escape degradation in the rumen may be digested as they pass further down the digestive tract either by animal enzymes secreted into the small intestine or by microbial enzymes in the large intestine. Since mammals do not produce enzymes capable of digesting structural carbohydrates, once they pass from the rumen, the only remaining, but limited, opportunity for further digestion of these plant components is in the large intestine.

Protein digestion

Proteins, as described earlier, contain carbon, hydrogen, oxygen, nitrogen, usually sulfur and sometimes phosphorus. In order to understand protein digestion, a brief description of their chemistry is required.

Amino acids

Each different protein consists of a unique combination of 20 different amino acids, each amino acid containing an amino group—a single atom of nitrogen combined with two atoms of hydrogen (NH_2).

Not all proteins contain all amino acids. These concepts are illustrated in figure 4.11.

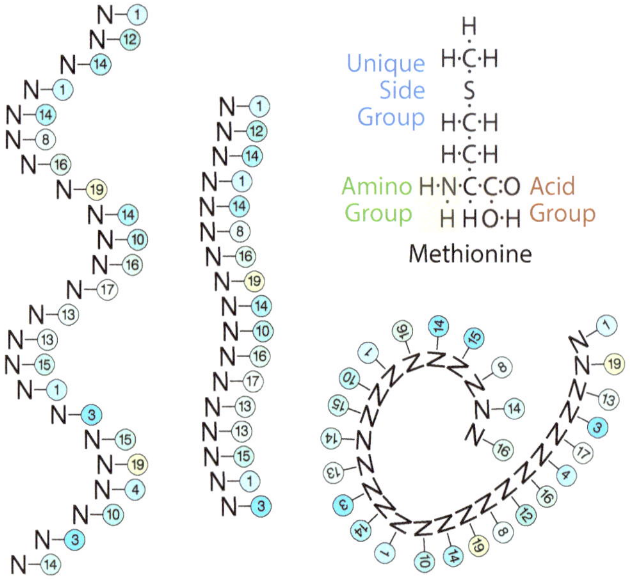

Figure 4.11: There are 20 different amino acids, each characterized by its unique side group (methionine illustrated here). Proteins are composed of long chains of up to thousands of amino acids, each different protein characterized by its unique amino acid sequence and shape.

Monogastric protein digestion

When protein is digested by the monogastric animal (figure 4.12), the long chains are first broken down into shorter chains called peptides, a process which begins in the stomach. In the small intestine, further digestion

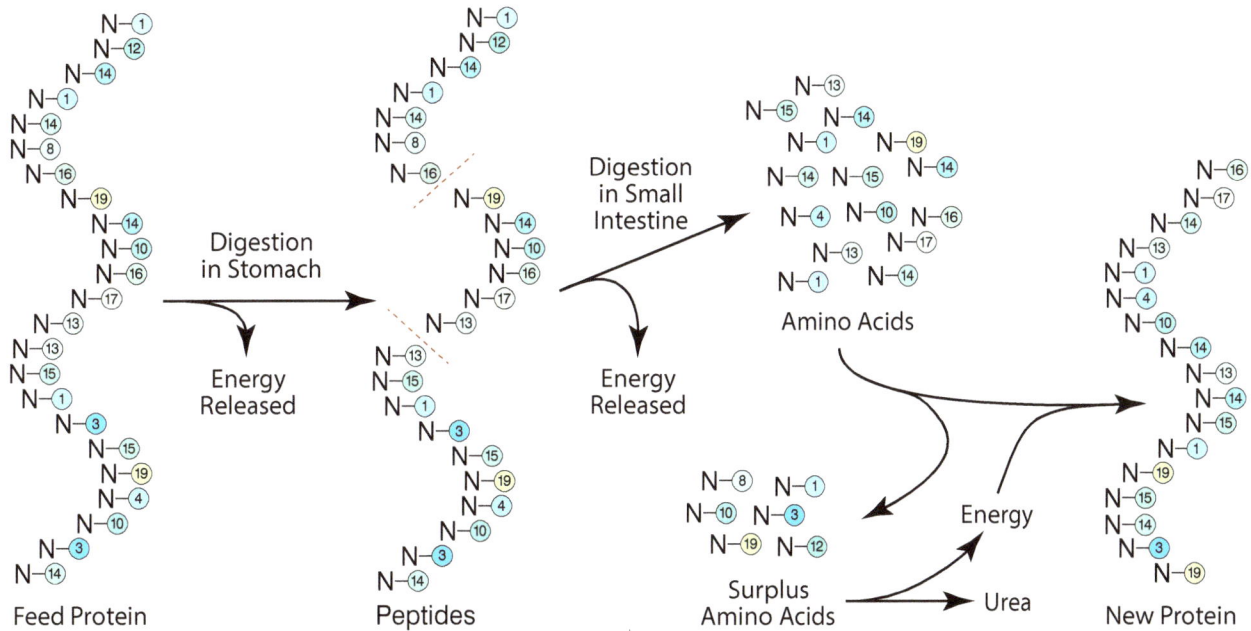

Figure 4.12: Protein digestion in the monogastric animal.

releases the individual amino acids which are absorbed into the bloodstream. The animal now uses these absorbed amino acids as building blocks for its own particular proteins.

Protein quality

Because animals have specific requirements for individual amino acids, the concept of protein quality arises. If the balance of amino acids in the feed protein is very similar to that required by the animal, the protein is said to be of high quality (table 4.2).

If the amino acid composition of feed proteins is poorly matched to requirements, the protein is of low quality. Surplus amino acids are broken down in the liver and kidney. The amino group is removed and may be recycled or excreted in the form of urea while the remainder of the molecule is used as an energy source.

Ten of the 20 amino acids can be synthesized by mammalian tissues as long as a source of nitrogen is available. For this purpose, nitrogen may be obtained from

Protein Source	Protein Quality
Grass Forage	38
Legume Forage	52
Corn Grain	58
Wheat Grain	59
Canola Meal	62
Mixed Microbes	65
Soyameal	70
Oats Grain	70
Barley Grain	72
Fishmeal	77
Milk	87
Egg	99

Table 4.2: Protein quality expressed as relative biological value.

urea or from the amino groups of surplus amino acids. The remaining 10 essential amino acids (page 57) must be supplied in the diet of the monogastric animal. It is the limited availability of these essential amino acids which usually restricts the synthesis of animal proteins. For example, when corn is fed to pigs, the limited availability of the essential amino acid lysine usually restricts growth. When corn-based rations are supplemented with lysine, a significant improvement in feed conversion efficiency occurs (see page 82).

Ruminant protein digestion

Protein digestion in the ruminant is much more complex than that in the monogastric. Feed proteins enter the rumen and reticulum where they are attacked by the microbial population and broken down to their constituent amino acids. However, microbial degradation does not stop here. Most of these amino acids are further degraded, resulting in the release of the amino group which, when combined with a third hydrogen atom, forms ammonia (NH_3) as shown in figure 4.13. As was the case in carbohydrate digestion, protein degradation releases energy.

Ammonia in the rumen can also be derived from non-protein nitrogen (NPN) sources. Mention was made earlier of the large quantities of saliva produced by the cow. This saliva serves as a vehicle for recycling amino groups in the form of urea (figure 4.14) back into the digestive system from amino acid breakdown elsewhere (e.g., liver). Urea may also be included in the diet as an inexpensive source of crude protein. Urea is broken down in the rumen by the enzyme urease and its amino groups are released as ammonia.

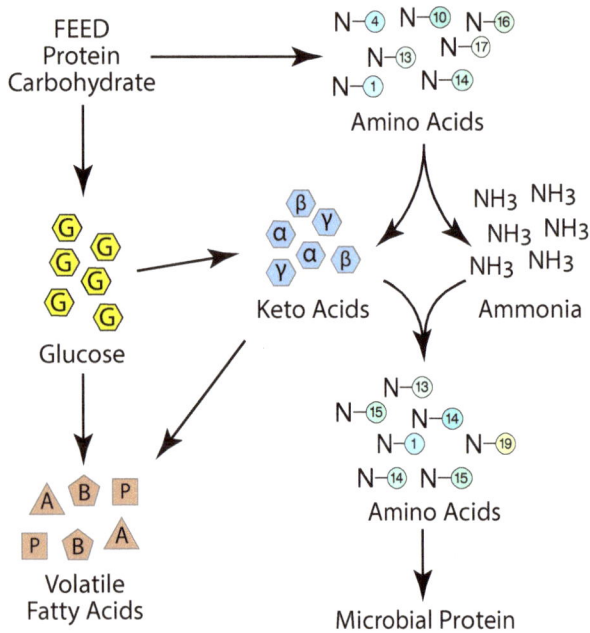

Figure 4.13: Protein metabolism in the rumen.

Reference has already been made to the fact that the bacterial population adherent to the rumen papillae is a major contributor of urease (page 44). The practical implications of this will be discussed later.

The first stage of protein digestion in the ruminant is completed with the production of amino acids and ammonia, accompanied by the release of energy.

The second stage involves the use of these products by the bacteria, protozoa and fungi in the rumen to build new microbial protein. This process, illustrated in figures 4.13 and 4.14, requires recapturing some of the energy released in the first stage as well as energy released from carbohydrate breakdown. The keto acids released during both protein and carbohydrate digestion are combined with ammonia to form new amino acids and, subsequently, new microbial protein.

Having completed the processes of plant protein degradation and microbial protein synthesis, micro-organisms are drawn through the omasum and into the abomasum. Here the digestion of microbial protein begins in a manner similar to that in the monogastric. Digestion continues in the small intestine with amino acids being absorbed into the bloodstream to provide building blocks for animal proteins.

Since the microbial population in the rumen has the capacity to synthesize essential amino acids, proteins that are of lower quality for monogastrics may be improved by the rumen microbes. On the other hand, higher quality proteins are also modified with the result that the protein reaching the small intestine is of a relatively uniform medium quality irrespective of the protein in the feed unless a significant proportion of rumen undegradable protein has been provided.

As described above, protein digestion in the ruminant involves two extra steps compared with that in the monogastric. Each extra step in the digestive process

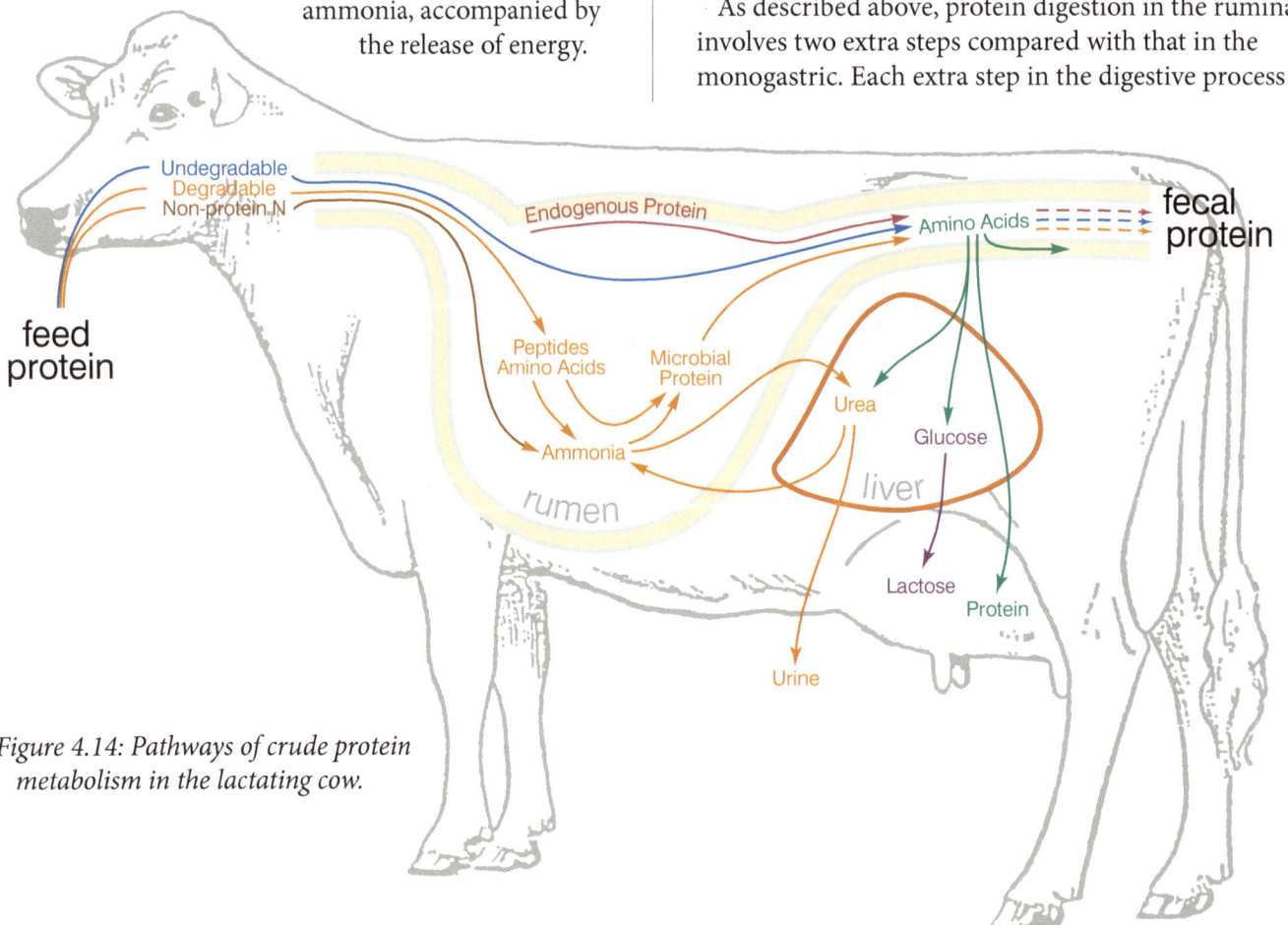

Figure 4.14: Pathways of crude protein metabolism in the lactating cow.

results in some loss of overall efficiency. This, combined with the fact that microbial protein is of mediocre quality, means that cattle are inefficient in utilizing high quality protein in comparison to monogastrics. Since high quality protein sources are usually expensive, these observations reinforce the concept that cattle rations must be based on inexpensive ingredients, primarily forages.

Urea in cattle diets

As mentioned earlier, urea appears in the rumen in association with saliva. Urea and other non-protein nitrogen (NPN) sources may also be added to cattle rations to increase the level of crude protein. However, to make efficient use of urea it is necessary to appreciate a few features of its metabolism.

Urea and other sources of NPN added to cattle rations are not efficiently utilized until the rumen microbial population becomes adapted to their increased availability. Although this process begins within a few days of the introduction of additional NPN, it may take several weeks before maximum utilization is attained. Therefore, short-term feeding of NPN supplements makes little sense.

The micro-organisms that produce urease (the enzyme responsible for urea breakdown) are concentrated on the inner lining of the rumen. Consequently, when large quantities of urea are fed over short periods of time, high concentrations of ammonia can accumulate near the rumen walls. This ammonia can produce a rapid increase in rumen pH and, after passing into the blood stream, can cause alkalosis (high blood pH). The effect is opposite to the effect produced by grain overload.

In order to be incorporated into microbial protein, the ammonia produced by urea degradation must be combined with organic keto acids to form amino acids (figure 4.13). In addition, protein synthesis requires far more energy than is released by urea breakdown. It is important, therefore, when feeding urea to also provide readily fermentable carbohydrates as a source of both keto acids and energy. In fact, this principle applies to the efficient utilization of ammonia in the rumen irrespective of its source. Rations should contain approximately 4.4 Mcal of digestible energy (DE) for every 120 grams of crude protein degraded in the rumen (rumen degradable protein; RDP).

When ammonia is produced in excess of the availability of keto acids and energy, it is absorbed through the walls of the rumen into the bloodstream. To some extent it may be recycled into saliva in the form of urea, some will appear in milk, but the greater proportion will simply be excreted (figure 4.14).

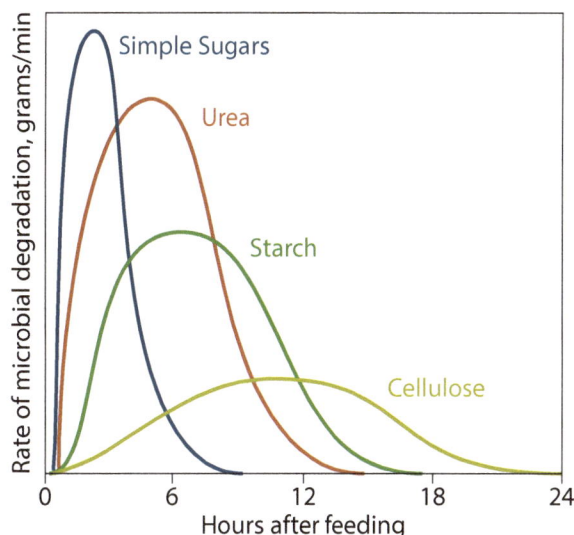

Figure 4.15: *Rumen microbial degradation rate of urea compared with rates for various carbohydrates.*

When animals are fed on a periodic basis (i.e., not self-fed *ad libitum*), it is important that the rate of carbohydrate degradation be well matched to the rate of urea degradation. Cellulose, for example, is inappropriate for this purpose because of its slow rate of digestion whereas the starch found in feed grains is ideal (figure 4.15). *Ad libitum* self-feeding, where feed is consumed in frequent, smaller meals results in more stable conditions in the rumen and these considerations become somewhat less important.

Rumen undegradable protein

It was suggested earlier that much of the feed protein entering the rumen is degraded to ammonia. In fact, the degree of degradation varies, depending on the source of protein (table 4.3, page 52). For most of the grass and legume forages, protein degradability (rumen degradable protein as % of crude protein) is in the 70 to 80% range; degradability of oats and barley protein is in the same range. Among the crude protein supplements, urea is considered 100% degradable; canola meal protein is 64 to 72% degradable while soymeal protein degradability is highly variable, depending on how it is processed. Fish meal protein is the extreme at approximately 22 to 26% degradability.

Protein that is not degraded in the rumen is termed rumen undegradable protein (RUP). Under some circumstances it is possible to use RUP to improve the overall quality of protein reaching the small intestine. Amino acids released from the RUP upon digestion there may complement the amino acids released from microbial protein, resulting in a better balance of amino acids being absorbed into the bloodstream. This can result in more efficient overall utilization of feed protein.

Protein Source	RDP, % of CP
Grass Hay	70–78
Grass Silage	74–82
Legume Hay	73-79
Barley Straw	53–60
Oats Grain	80–85
Barley Grain	74–80
Corn Grain	53–61
Soymeal	37-67
Fish meal	22-26
Corn Distillers Grains	53-58
Corn Gluten Meal	30–35
Canola Meal	64–72
Urea	100

Table 4.3: Rumen degradable protein (RDP) as % of crude protein (CP) in some common western Canadian dairy ration ingredients.

The concept of protein degradability explains why some feeds yield better performance results than others although the total amount of crude protein provided in the ration is the same. For example, production is often improved when soymeal or fish meal is used in place of lower quality protein sources. In a summary of trials where various sources of RUP were included in rations for lactating dairy cows, fish meal consistently produced the highest milk yields. This was likely due to the RUP in fish meal providing a better complement of amino acids than those provided by microbial protein (see Protein quality, page 49).

Heat-damaged protein

When silage is poorly packed, excess air can result in significant heating. Similarly, hay that is baled when its moisture content is too high may produce heat. Heating is caused by chemical reactions (*Maillard* also called *browning* reactions) in the feed which result in specific protein amino acids becoming bound with simple sugars. The compounds formed are less digestible by rumen microbes. When heat damage is slight, these compounds may be digestible in the abomasum and small intestine serving a role similar to that of rumen undegradable protein. However, in many cases, heat-damaged protein will be totally indigestible, with the amount of loss being proportional to the degree of heating. A chemical test (acid detergent fibre nitrogen) to determine the degree of loss was discussed in the section on Feed Analysis (pages 33 - 34).

Lipid Digestion

Lipids, commonly referred to as fats or oils, are primarily composed of fatty acids (FA), appearing in both plants and animals as either 'free' FA (FFA; also called 'non-esterified' FA (NEFA)) or linked (esterified) to a glycerol molecule to form mono-, di- or triglycerides. Other classes of lipids include phospholipids, galactolipids and steroids but these are of minor nutritional significance.

Fatty acids are composed of chains of carbon atoms, ranging in length from 2 to 22+ carbons. The volatile fatty acids (VFA) produced by rumen fermentation are 2 to 6 carbon atoms in length. At the opposite end of the spectrum are the nutritionally important 20 and 22 carbon omega FA found in marine organisms,

The first carbon atom in the FA carbon chain carries an organic acid group. When all of the remaining carbon atoms in the chain carry the maximum number of hydrogen atoms, the FA is said to be 'saturated'. A 'mono-unsaturated' FA has a pair of hydrogen atoms missing from each of 2 adjacent carbon atoms. A 'polyunsaturated' FA has more than one missing pair of hydrogen atoms on carbon atoms regularly spaced 2 carbon atoms apart.

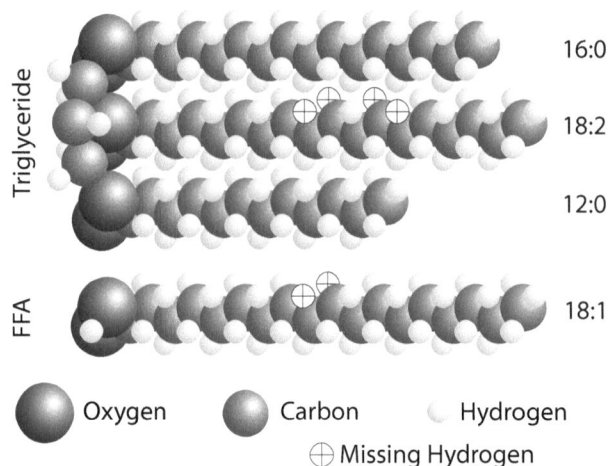

Figure 4.16: Lipids are composed of fatty acids (FA) in free form (FFA) and as mono-, di- and triglycerides. The FA 'backbone' is a chain of 2 - 22+ carbon atoms. A saturated FA has 2 hydrogen (H) atoms attached to each carbon (the terminal carbon always has 3). When a single pair of H atoms are missing from adjacent carbons, the FA is monounsaturated (MUFA); when more than one pair are missing it is termed polyunsaturated (PUFA). In the diagram above, 16:0 is a 16-carbon (C16) FA with zero H atoms missing—it is saturated; 18:1 is a monounsaturated C18 FA with 1 pair of H atoms missing; 18:2 is a polyunsaturated C18 FA with 2 pairs of H atoms missing.

52

Lipid intake

The range of lipid concentrations in common base ingredients used in typical western Canadian dairy rations is shown in table 4.4. A balanced ration composed of these ingredients will result in basal dietary crude fat concentrations in the range of 3 - 4% of diet dry matter (DM). Supplementary lipid sources (see page 73) aimed at increasing dietary energy concentration may add up to an additional 2 to 3 percentage points of crude fat.

Ration Ingredient	Crude Fat % of DM
Grass Hay	2.8 - 4.0
Legume Hay	2.1 - 3.0
Grass Silage	2.9 - 4.4
Barley Silage	2.6 - 3.7
Corn Silage	2.5 - 3.3
Oats Grain	4.3 - 7.1
Barley Grain	2.0 - 2.6
Corn Grain	3.4 - 4.3
Soymeal, Solvent Extracted	1.0 - 2.7
Canola Meal	2.7 - 4.4
Distillers Grains with Solubles	5.6 - 14.2
Corn Gluten Meal	1.4 - 3.5
Fishmeal	8.6 - 12.3
Beet Pulp	0.8 - 1.4

Table 4.4: Typical crude fat concentrations in western Canadian dairy ration base ingredients.

The primary fatty acids provided by common dietary ingredients are 16 and 18 carbon long-chain FA (LCFA) (see Appendix table A2.5), including:
• palmitic acid (C16:0);
• stearic acid (18:0);
• oleic acid (C18:1);
• linoleic acid (C18:2), and;
• alpha linolenic acid (C18:3).

Lipid metabolism in the rumen

Dietary mono, di- and triglycerides are rapidly hydrolyzed by rumen microbes to yield free fatty acids and glycerol, a 3-carbon monosaccharide (sugar). Following hydrolysis, long chain (C16 - 18) unsaturated FAs can be *biohydrogenated* by ruminal bacteria to produce saturated FA of equivalent carbon chain length. Complete biohydrogenation of these unsaturated FA involves several steps that may be performed by different microbial species. Therefore, the extent of biohydrogenation and the products of hydrogenation vary, depending on microbial access to the FA and their rate of passage through the rumen. Estimates of the extent of biohydrogenation of polyunsaturated FA (PUFA) entering the rumen have ranged from 60 to 90%.

Very little oxidation of LCFA occurs in the rumen although FA with chain lengths of 14 or less may be oxidized. Rumen microbes may also synthesize FA with chain lengths of 8 to 14 carbon atoms and FA may be absorbed by the rumen epithelium. The net result is that, without supplementation of 'bypass fats', the amount of FA reaching the small intestine is normally around 80 to 84% of that entering the rumen. Palmitic, stearic and oleic acids are the main FA that ultimately pass from the rumen. Because FAs are hydrophobic (water-repellent), they pass through the omasum and abomasum in association with feed particles.

Polyunsaturated FA as well as saturated C12 and C14 FAs (lauric and myristic acids) have been reported to have negative effects on rumen fibre digestion. In one study, supplementing vegetable oils at 3% of diet DM decreased NDF digestibility by about 1.3 percentage units while other studies have found this effect to be negligible. It is generally recommended that lipid supplements high in PUFA should be limited to 1.5 to 2% of diet DM. In contrast, a number of studies have demonstrated improved NDF digestibility with supplements containing higher levels of palmitic acid (C16:0).

Lipid metabolism in the small intestine

On reaching the small intestine, FFAs are mixed with bile salts and pancreatic secretions. Once they have been thoroughly emulsified, the FAs are absorbed into the lymphatic system then slowly released into to the bloodstream for delivery to muscles, adipose (fat) tissues, the liver and the mammary gland.

When diets are supplemented with lipids in the form of triglycerides (TG) that are protected from rumen hydrolysis and biohydrogenation (bypass fats), the TG must be hydrolyzed in the small intestine. Bile acids emulsify the lipids and TG-hydrolyzing pancreatic enzymes (lipases) free the FAs from the TG glycerol backbone. While moderate amounts of TG-containing bypass fats are emulsified and hydrolyzed efficiently, the cow's capacity to do so is limited by the amount of pancreatic lipase that can be produced. Therefore, it is recommended that bypass fat supplementation be limited to 1.5 - 2% of diet DM.

Chapter 5: Nutrients

Water

Water is the most important nutrient for dairy cattle. It is required for all of life's processes, including:

- transport of nutrients and other compounds to and from cells;
- digestion and metabolism of nutrients;
- elimination of waste materials as urine, feces and respiratory water vapour;
- maintenance of body temperature due to its high heat capacity and removal of excess heat from the body through evaporation;
- lubrication of joints and other organs;
- maintenance of fluid and ion balance;
- provision of a fluid environment for the developing fetus; and
- milk production.

The total body water content of dairy cattle is 56 to 81 percent of body weight, depending on physiologic stage and body composition. Cows in early lactation have more body water (69% of body weight) than cows in late lactation (62%). Late gestation dry cows are intermediate in body water content (65%). Fat cows have lower body water content than thin lactating cows and younger, leaner animals have a higher water content than older animals.

Body water is divided into intracellular and extracellular compartments. Intracellular (within the cell) water is the largest compartment, accounting for about 40% of body weight and two-thirds of the water in the body. The extracellular (outside the cell) fluid includes water around cells and tissues, water in plasma, and water in the gastrointestinal tract. Intestinal water accounts for 10-35 percent of body weight—in early lactation cows it is about 15%; in pregnant and late lactation cows it is around 10-11%.

Dairy cattle derive water from feed, from metabolic oxidation reactions and from the consumption of 'free' water. If free water is readily available throughout the day, the volume of water in the body remains relatively constant. Water is lost via feces, urine, sweating and respiration, while a small volume is also lost through saliva; milk is a significant route of water loss in lactating cows.

Lactating cows should have access to unlimited quantities of clean, high quality water throughout the day. Water quality should be routinely evaluated by chemical and microbiological analysis of samples from a convenient access point, typically in the milk house. Table 5.1 suggests maximum concentrations of various ions in free water offered to cattle.

Many well waters in western Canada contain high levels of iron and sulfates, facilitating the growth of iron- and sulfur-utilizing bacteria that can clog pipes and valves and impart a 'rotten egg' odour to the water due to the production of hydrogen sulfide. High nitrate and fecal bacteria levels can result from contamination of well water by manure.

Although lab testing provides information about the quality of water close to its source, that quality can be negatively affected by fecal contamination and by the growth of bacteria and algae in waterers. For this reason, waterers should be routinely cleaned, including their float and heater compartments. 'Tip trough' waterers like that shown in figure 5.1 provide the large capacity preferred by cows along with ease of cleaning.

	Maximum Recommended Concentration (mg/L)
Major Ions	
Calcium	1,000
Nitrate and nitrite	100
Nitrite alone	10
Sulfate	1,000
Total Dissolved Solids	3,000
Heavy Metals and Trace Ions	
Aluminum	5.0
Arsenic	0.2
Beryllium	0.1
Boron	5.0
Cadmium	0.02
Chromium	1.0
Cobalt	1.0
Copper	1.5
Fluoride	2.0
Iron	0.3
Lead	0.1
Manganese	0.05
Mercury	0.003
Molybdenum	0.5
Nickel	1.0
Selenium	0.05
Uranium	0.2
Vanadium	0.1
Zinc	50.0

Table 5.1: Maximum recommended concentrations of ions in free (drinking) water offered to cattle.
sources: Canadian Water Quality Guidelines, 1987; NASEM Dairy 8, 2021.

Figure 5.1: A 'tip trough' waterer combines high capacity with ease of cleaning.

Factors that affect the pattern and amount of free water intake include:

- water accessibility and quality;
- feeding and milking times—cows will normally consume around 40% of their daily water intake within 2 hours of each event;
- social standing—dominant cows drink more than recessive herdmates;
- waterer size—cows consume more water from larger, deeper troughs;
- moisture content of feed—animals will consume less free water as the proportion of wet ingredients (e.g., silage) in the diet increases.

Several equations have been proposed for the prediction of free water intake (FWI) by cows. Those recommended in the 2021 NASEM Dairy 8 publication are as follows:

For lactating cows when a reliable estimate of dry matter intake is available:

FWI, kg/day = –91.1
  + 2.93 x Dry Matter (DM) Intake, kg/day
  + 0.61 x Dietary DM %
  + 0.062 x sum of Na and K concentrations in the diet, milliequivalent/kg of DM:
    (% Na/0.023) + (% K/0.039) x 10
  + 2.49 x Dietary Crude Protein %
  + 0.76 x Daily Mean Ambient Temperature, °C

For lactating cows when a reliable estimate of dry matter intake is not available:

FWI, kg/day = –60.2
  + 1.43 x Daily Milk Yield, kg/day
  + 0.064 x sum of Na and K concentrations in the diet, milliequivalent/kg of DM:
    (% Na/0.023) + (% K/0.039) x 10
  + 0.83 x Dietary DM %
  + 0.54 x Daily Mean Ambient Temperature, °C
  + 0.08 x Days in Milk

For dry cows:

FWI, kg/day = 1.16 x Dry Matter (DM) Intake, kg/day
  + 0.23 x Dietary DM %
  + 0.44 x Daily Mean Ambient Temperature, °C (TMP)
  + 0.061 x $(TMP – 16.4)^2$

Energy

When carbohydrates, lipids and proteins are broken down, energy is released. While a portion of the released energy is lost as heat, much of the remainder is captured, to be used in the growth and maintenance of body tissues, in conceptus growth and in the synthesis of milk.

Figure 5.2 illustrates the loss and capture of energy as feed is digested and the products of digestion are used by the animal. Gross energy (GE) is estimated by burning a sample of feed in an atmosphere of oxygen and measuring the amount of heat produced (see page 37). Some of this is unavailable to the animal because it is indigestible, passing through the entire digestive tract to be excreted as feces. Of the energy released by digestion (digestible energy, DE), portions are lost from both ends of the digestive tract in the form of methane, hydrogen and other gases while another portion is lost through urinary excretion.

Metabolizable energy (ME) is available to the animal for maintenance, growth, reproduction and milk production. These processes result in the loss of heat (heat of production) which, when ambient temperature

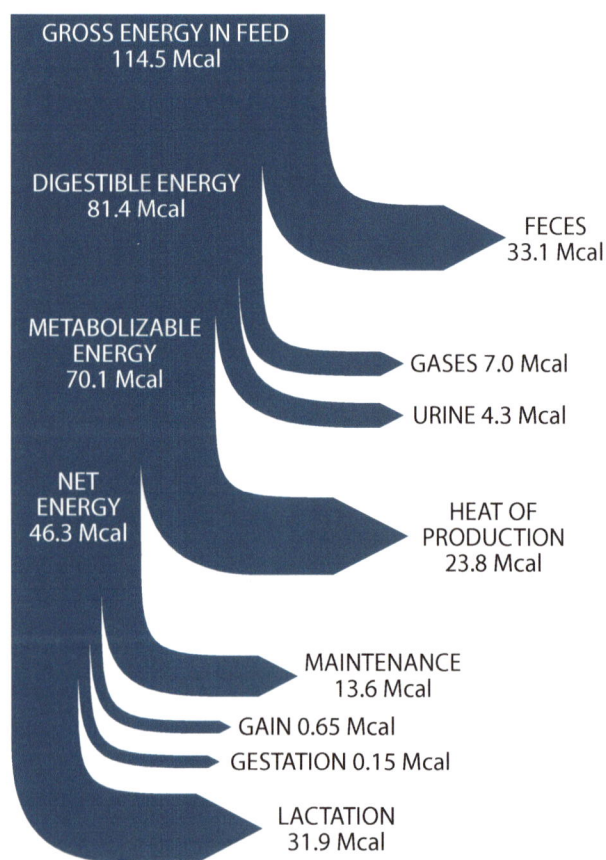

Figure 5.2: Partitioning of energy in a 3rd lactation cow, weighing 700 kg, producing 42 kg/day of 3.8% bf milk, gaining 0.13 kg/day at 200 days in milk, 100 days pregnant.

is low, serves to keep the body warm. The remaining net energy represents the amount that is actually captured by processes which require the synthesis of new organic compounds—carbohydrates, lipids, proteins and nucleic acids. When describing feed energy values and animal energy requirements, net energy is identified by the specific process it supports: Net Energy for Maintenance (NEM), Net Energy for Gain (NEG), Net Energy for Lactation (NEL), or Net Energy for Gestation.

The efficiency with which ME is typically used for milk production (i.e., available as NEL; 64‑66%) is similar to the efficiency of ME use for maintenance (i.e., available as NEM). Therefore, the 2021 NASEM Dairy 8 guidelines use NEL to define requirements for both lactation and maintenance as well as for pregnancy, growth and changes in body reserves of adult cows. For heifers prior to their first calving, requirements are stated on an ME basis.

Because the yield of ME and NE for each dietary ingredient can be influenced by the other ingredients, ME and NE values are not valid for individual feeds. Therefore, feed energy values in Appendix A are stated in terms of 'DE Base' values which are computed from the organic matter composition of each feed.

Protein

Dairy cattle manufacture a vast array of different proteins, ranging from milk casein to actin in muscle, hemoglobin which carries oxygen in blood and pepsin which digests dietary protein in the abomasum. And virtually all biochemical reactions are catalyzed by proteins acting as enzymes. To synthesize these proteins, the animal requires amino acids, the basic subunits of all proteins. Of the 20 amino acids commonly found in mammalian proteins (table 5.2), nine are considered essential for non-ruminant mammals—they must be provided in the diet as they cannot be manufactured by mammalian tissues. Another six amino acids are considered conditionally essential for non-ruminants; these may be required to some degree in young, growing animals and/or during illness. The remaining five amino acids are classified as non-essential in the diet because they can be synthesized by mammalian tissues.

Ruminant animals do not require dietary amino acids or true protein in their diets to support maintenance or very low levels of production. This is because the animals' amino acid requirements are satisfied by digesting microbial protein that passes down the digestive tract from the rumen. Rumen microbes are able to manufacture all of the amino acids from dietary rumen degradable protein (RDP) and non-protein nitrogen (NPN) sources that yield ammonia after entering the rumen. To support anything beyond very

Class	Amino Acid	Microbial	Empty Body	Milk
		grams AA / 100 grams of protein		
Essential Amino Acids	Histidine	2.21	3.04	2.92
	Isoleucine	6.99	3.69	6.18
	Leucine	9.23	8.27	10.56
	Lysine	9.44	7.90	8.82
	Methionine	2.63	2.37	3.03
	Phenylalanine	6.30	4.41	5.26
	Threonine	6.23	4.84	4.62
	Tryptophan	1.37	1.05	1.65
	Valine	6.88	5.15	6.90
Conditionally Essential	Arginine	5.47	8.20	3.74
	Cysteine	2.09	1.74	0.93
	GlutamX[1]	14.98	15.76	22.55
	Glycine	6.26	14.46	2.04
	Proline	4.27	9.80	10.33
	Tyrosine	5.94	3.08	5.83
Non Essential	Alanine	7.38	8.59	3.59
	AsparX[2]	13.39	9.61	8.14
	Serine	5.40	5.73	6.71

Table 5.2: Amino acids classified by their essentiality to mammalian metabolism and their contributions to microbial, empty body and milk proteins. [1]GlutamX = Glutamate (non-essential) + Glutamine (conditionally essential); [2]AsparX = Aspartate + Asparagine (both non-essential).

low production, a few of the essential amino acids are required in the diet to support the efficient processing of forages by rumen microbes.

At the very high levels of production expected of the lactating dairy cow, high quality protein which is resistant to microbial degradation must also be fed. Such rumen undegradable protein (RUP) must be digestible in the small intestine and must supply a mixture of amino acids which complements the amino acids provided by microbial protein (see page 51).

Because RUP sources are generally more expensive than RDP, and because microbial protein provides the greater proportion of absorbable amino acids, it is important to maximize microbial protein production by providing rumen microbes with the RDP and energy-yielding substrates required for this purpose before considering the addition of RUP.

In the NASEM Dairy 8 publication, protein nutrition is described in terms of metabolizable protein, partitioned as illustrated in figure 5.3. However, it must be recognized that it is the individual amino acids (AA)

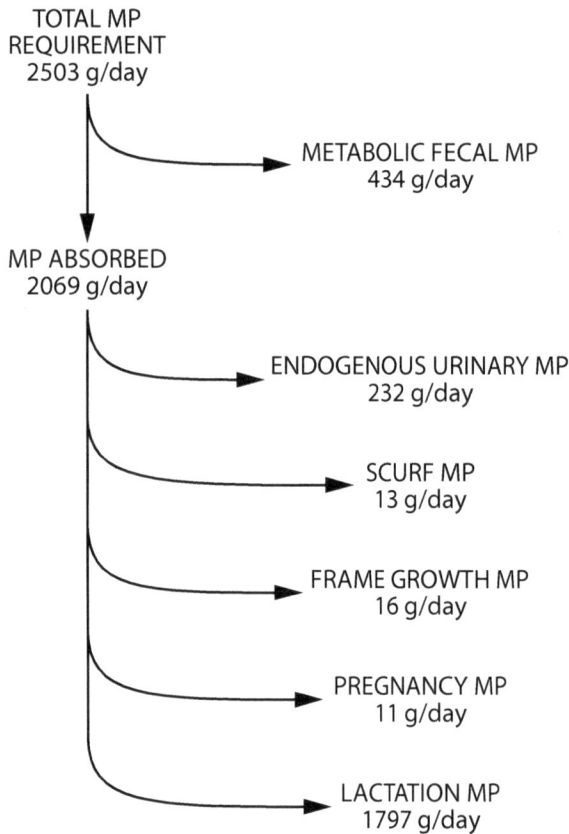

Figure 5.3: Metabolizable Protein (MP) utilization in a 4.5 year old Holstein cow weighing 700 kg, 200 days in milk, 100 days pregnant.

that are absorbed from the digestive tract and utilized in various metabolic processes rather than protein *per se*. Ideally, protein requirements for these processes should be stated in terms of the unique blend of AA required. For example, milk protein synthesis requires a blend of AA that is different from that required for conceptus development. While knowledge in this area has increased considerably over recent decades, it is not yet possible to define protein requirements in terms of quantities of absorbed amino acids. Until that becomes possible, recommendations will be defined in terms of metabolizable protein (MP), recognizing that, although a formulation may contain adequate MP, it may not contain an optimum blend of essential AA.

The blend of amino acids available for absorption in the cow's small intestine is a combination of those released by intestinal digestion of both microbial protein and digestible rumen undegradable protein. This is referred to in figure 5.3 as MP Absorbed. Another portion of the true protein (i.e., not including NPN) traversing the digestive tract will be indigestible and will be excreted in the feces along with endogenous protein resulting from digestive tract tissue turnover. This fraction is illustrated in figure 5.3 as metabolic fecal MP.

Although the primary role of absorbed amino acids is in the synthesis of proteins, they are also involved in many other metabolic functions. For example, because ruminants absorb very little glucose from their digestive tracts, glucose must be synthesized in the liver from other metabolites, mainly propionic acid (one of the volatile fatty acids produced by microbial fermentation) and amino acids. The synthesis and secretion of lactose by the mammary gland in lactating cows places a very significant demand on this synthesis of new glucose (gluconeogenesis). Another major fraction of absorbed amino acids is required for tissue turnover, a process by which tissues are continually replaced by new tissue. These are accounted for in the MP scheme (figure 5.3) as scurf MP and frame growth MP. Amino acids can also be converted to fatty acids or serve as a source of energy, particularly when the balance of amino acids available for metabolic processes is out of balance with requirements. Oxidation of amino acids yields energy and ammonia which is converted to urea by the liver and may be recycled back to the rumen or excreted as endogenous urinary MP.

Minerals

Fifteen minerals have been demonstrated as essential in dairy nutrition. Of these, seven are termed macrominerals, because they are required in relatively large quantities. The remaining eight essential minerals are referred to as trace or microminerals. Although these are required in relatively small quantities, it should not be inferred that they are any less important. The essential minerals are listed in table 5.3.

Table 5.4 describes the functions and signs of deficiency of the essential minerals. Notice in the column entitled 'Special Considerations' that reference is made to interactions between minerals and between specific minerals and vitamins. Some of these demand particular attention because they are often involved in practical feeding problems.

Macrominerals	Micro (trace) minerals
Calcium (Ca)	Cobalt (Co)
Chloride (Cl)	Copper (Cu)
Magnesium (Mg)	Iodine (I)
Phosphorus (P)	Iron (Fe)
Potassium (K)	Manganese (Mn)
Sodium (Na)	Molybdenum (Mo)
Sulfur (S)	Selenium (Se)
	Zinc (Zn)

Table 5.3: Essential macro- and micro- (trace) minerals.

Macromineral	Primary Functions	Deficiency Signs	Special Considerations
Sodium (Na) and Chlorine (Cl); Salt (NaCl)	· Na modulates extracellular fluid volume and acid-base equilibrium · heart function and nerve impulse transmission (with K) · nutrient transport (Na-K pump) · salivary Na bicarbonate is an important rumen pH buffer	· chewing wood, licking dirt, eating toxic amounts of poisonous plants (pica) · decreased appetite and feed efficiency, weight loss, unthrifty appearance, rough hair coat · incoordination, shivering, weakness, dehydration, cardiac arrhythmia leading to death	· 30-50% of total body Na is stored in bone · nerve conduction dependent on Na/K balance · when mixed with rations or mineral mixes, may be used to limit their consumption · low Na intake increases K concentration in rumen fluid
Calcium (Ca)	· formation of skeletal tissues · transmission of nerve impulses · excitation of skeletal and cardiac muscle contraction, · blood clotting · milk synthesis	· in young animals: abnormal development of bone and, in severe cases, rickets and tetany · in older animals: osteoporosis and osteomalacia due to withdrawal from bone	· Ca deficiency symptoms develop slowly as calcium is drawn from bone · excessive Ca can interfere with absorption of Zn and Se · vitamin D required for proper utilization of both Ca and P · Ca:P ratio should be in 1:1 to 5:1 range · Ca and P concentrations in milk are not altered even during severe dietary deficiency · mature forages are often low in P
Phosphorus (P)	· formation of bones and teeth · regulation of enzyme activity through phosphorylation · energy metabolism (ATP/ADP) · acid-base balance of blood and other bodily fluids · a component of cell membranes and cell contents as phospholipids and nucleic acids	· slow growth · loss of or depraved appetite · unthrifty appearance · poor reproductive rates · reduced milk yield · hemoglobinuria and liver dysfunction · symptoms similar to those due to Ca deficiency	
Magnesium (Mg)	· a cofactor for many enzymes · involved in nerve conduction, muscle function, bone mineral formation, Ca and P homeostasis	· irritability, muscle twitching, hyperexcitability, convulsions · tetany (most often with cows on fast growing pasture in the form of grass tetany)	· function and metabolism of Mg closely tied to that of Ca and P · high dietary K reduces Mg absorption
Potassium (K)	· osmotic pressure and acid-base regulation, water balance, nerve transmission, muscle contraction, oxygen and carbon dioxide transport · an activator or cofactor for many enzymes involved in protein synthesis and carbohydrate metabolism	· marked reduction in feed and water intake, pica, loss of body weight and milk yield · dull hair coat, decreased pliability of the hide · severely deficient cows will be profoundly weak or recumbent	· deficiency most common when high concentrate diets are fed · higher-yielding cows affected more quickly and severely than lower-yielding cows · low Na intake increases K concentration in rumen fluid
Sulfur (S)	· a component of thiamin and biotin, the essential amino acids methionine and cysteine and other important molecules · sulfate ion (SO_4^{-}) is important in acid-base balance	· similar to protein deficiency · excess salivation, runny eyes, hair shedding	· dietary nitrogen (CP/6.25): sulfur ratio should be in 10:1 to 12:1 range · diets high in urea may be low in sulfur

Table 5.4a: Essential macrominerals; their functions, deficiency and toxicity symptoms and special considerations relating to nutrient management.

Micromineral	Primary Functions	Deficiency Signs	Special Considerations
Cobalt (Co)	· cofactor in vitamin B12 synthesis by rumen microbes	· deficiency symptoms are mainly due to inadequate vitamin B12 synthesis · unthrifty appearance, anemia, emaciation and weakness · decreased growth, milk production and fertility	· should always be present in salt and mineral included in concentrate mixes · vitamin B12 is an essential cofactor in gluconeogenesis (synthesis of glucose from propionate and amino acids)
Copper (Cu)	· a component of several important enzymes · required for hemoglobin synthesis · related to absorption of iron · along with Zn, a component of an enzyme that protects cells from oxidative damage	· impaired immune function in cows which may lead to more severe cases of mastitis · loss of hair or hair pigment · anemia, reduced growth rate, increased prevalence of disease, reduced reproductive efficiency	· copper requirement depends on molybdenum level in diet—Cu:Mo ratio should be 5:1 · Cu consumed at high levels over long periods accumulates in the liver—stress can result in rapid release, jaundice and death
Iodine (I)	· formation of triiodothyronine (T3) and thyroxine (T4) in the thyroid gland, hormones that regulate energy metabolism	· enlargement of the thyroid gland (goiter) in newborn calves and adult cattle · calves may be born hairless, weak or dead · reduced fertility in cows	· deficiency common when non-iodized salt or mineral is fed · excess dietary iodine excreted in milk may result in milk levels exceeding regulatory limits
Iron (Fe)	· a component of hemoglobin, facilitating O_2 transport and several enzymes important in energy metabolism, immunity and gene regulation	· anemia in calves · deficiency is rare in adult cattle	· 'white' veal is due to low muscle myoglobin resulting from feeding a low iron diet
Manganese (Mn)	· a cofactor in a number of enzymes and other proteins required for metabolism of amino acids, carbohydrates and lipids and in the synthesis of cartilage and bone	· skeletal abnormalities in calves including enlarged joints and shortened and weak bones · equilibrium problems and ataxia can be caused by improper development of bones in the middle ear · poor reproductive efficiency	· exact requirements are not known with any certainty · currently no sensitive methods of assessing Mn status are available
Molybdenum (Mo)	· aids digestive process but interferes with Cu absorption	· high levels (>1 ppm) may provoke copper deficiency	· Cu:Mo ratio should be in 5:1 range
Selenium (Se)	· a component of antioxidant enzymes that also involve vitamin E	· nutritional muscular dystrophy (white muscle disease) · poor growth rate, impaired fertility, retained placenta	· Se deficiency is common in western Canada—in beef cattle it has been linked with high levels of calf loss
Zinc (Zn)	· a component of enzyme systems catalyzing nutrient metabolism, immune function, gene and hormonal regulation, and nerve transmission · structure of keratin proteins	· reduced dry matter intake and growth rate · flaky skin (parakeratosis) on legs, head, nostrils or neck · weak hoof horn	· when dietary concentration exceeds requirement, Zn homeostasis is maintained by fecal excretion of excess · healthy hooves require Zn

Table 5.4b: Essential microminerals; their functions, deficiency and toxicity symptoms and special considerations relating to nutrient management.

Mineral interactions

Dietary cation-anion difference (DCAD)

DCAD is the difference between the sum of the major cations (Na^+ and K^+) and the sum of the major anions (Cl^- and S^{2-}) in the diet, expressed in milliequivalents (mEq) per kg of diet dry matter. Research has suggested that a DCAD of -75 to -200 mEq/kg dry matter in the diets of late dry cows would reduce the risk of hypocalcemia (milk fever) at calving. The proposed mechanism involves negative DCAD lowering blood pH resulting in the mobilization of buffers including Ca phosphate and bicarbonate from bone. It is suggested that the increased Ca mobilization would help alleviate the potential Ca deficit for early lactation milk synthesis.

Calcium and phosphorus

These two minerals require adequate magnesium and vitamin D for efficient utilization. In addition, the ratio of calcium to phosphorus in rations should be in the 1:1 to 5:1 range. Higher proportions of calcium can reduce phosphorus absorption and, conversely, high phosphorus levels can reduce calcium availability.

Copper, molybdenum and sulfur

Many feeds grown in western Canada are deficient in copper. Furthermore, feeds which contain adequate copper levels in company with high molybdenum levels are effectively copper deficient. Excess molybdenum reduces copper availability and high sulfur levels amplify this effect.

Selenium and vitamin E

Both selenium and vitamin E are involved in the maintenance of tissue membranes; they have a 'sparing' effect on one another. Levels of selenium which are more than adequate can reduce the requirements for vitamin E and vice versa. There is no good evidence to suggest that either nutrient influences the absorption of the other. The range between selenium requirement and toxicity is particularly narrow; extra care must be taken not to oversupplement.

Cobalt and vitamin B12

Cobalt is an essential component of vitamin B12. When cobalt intake is inadequate, B12 synthesis by rumen microbes is reduced and symptoms of vitamin deficiency develop, including the reduced capacity of the liver to synthesize glucose from propionic acid and amino acids (gluconeogenesis), leading to decreased milk production.

Vitamins

Vitamins are small organic molecules that are essential co-factors in many enzyme-catalyzed metabolic pathways. They are commonly classified as either fat-soluble or water-soluble as in table 5.5.

Fat-soluble vitamins include A, D, E, and K. These vitamins are transported through the body in association with lipids and are absorbed by and stored in body fat. Vitamins A, D, and E are derived from the diet or administered by intramuscular or subcutaneous injection. Vitamin K is synthesized by microbes in the digestive tract.

Vitamin A refers collectively to a group of fat-soluble 'retinoids', including retinol, retinal, retinoic acid and dehydroretinol. Their names reflect the important role of vitamin A in the synthesis of the retinal pigment rhodopsin in the eye. These molecules do not occur in plants; however, plants contain carotenoids (yellow, orange, and red organic pigments) that are converted to vitamin A precursors (A1 and A2) by enzymes located in the intestinal epithelium of animals. Fresh, vegetative-stage forages are rich sources of carotenoids but their concentrations decline steadily after harvest and during storage.

Vitamin	Synonyms
Fat-soluble	
Vitamin A1	Retinol, retinal, retinoic acid
Vitamin A2	Dehydroretinol
Vitamin D2	Ergocalciferol
Vitamin D3	Cholecalciferol
Vitamin E	α-tocopherol, tocotrienols
Vitamin K1	Phylloquinone
Vitamin K2	Menaquinone
Vitamin K3	Menadione
Water-soluble	
Vitamin B1	Thiamin
Vitamin B2	Riboflavin
Vitamin B3	Nicotinamide, niacin
Vitamin B5	Pantothenic acid
Vitamin B6	Pyridoxol, pyridoxal, pyridoxamine
Vitamin B7	Biotin
Vitamin B9	Folacin, folic acid, folate
Vitamin B12	Cobalamin
Vitamin C	Ascorbic acid
Choline	Gossypine

Table 5.5: Vitamins required by all mammals.

Vitamin D3 is produced in some plants but the wide variability between plants and within plant species as they mature means that most feeds cannot be relied upon to supply animal requirements. Vitamin D2 is produced in the skin of mammals through a chemical transformation facilitated by exposure to ultraviolet radiation (i.e., sunlight). Dairy cattle housed indoors will not receive enough sunlight to satisfy their vitamin D requirements by this means so routine supplementation is required, except where grazing is practised in the summer months.

Vitamin E is a term that refers to a number of different compounds, the most biologically active being α-tocopherol. Fresh forages and oilseeds can contain appreciable concentrations of α-tocopherol as can dried distillers grains; whole grains and oilseed meals contain significantly less. However, like vitamin D2, vitamin E is subject to oxidization during storage and, therefore, the amount available from feedstuffs cannot be relied upon to satisfy requirements. Supplementation is required, either through the diet or by subcutaneous or intramuscular injection.

Although forages can be a source of vitamin K, cattle and other ruminants do not require a dietary source of this vitamin because sufficient quantities are synthesized by rumen microbes. As described on pages 20-21, apparent vitamin K deficiency may occur when mouldy hay is fed due to the activity of dicoumarol.

Pre-ruminant calves satisfy their fat-soluble vitamin requirements through colostrum, milk and/or milk replacer. See Appendix B table B5 for recommended milk replacer vitamin levels.

Feeds normally provide substantial amounts of the water-soluble vitamins and all can be synthesized either by rumen microbes or within body tissues, so they are not normally supplemented. However, they are not stored in the body and there is some evidence that supplementation of folic acid and vitamin B12 in the diets of high producing cows might be beneficial. Biotin (B7) supplementation has been shown to reduce the incidence of some types of hoof lesions and niacin (B3) is often included in transition diets for fat cows although evidence to support this use is inconclusive.

Chapter 6: Feeding Management

Calves and heifers: birth to calving

Growth targets

The ultimate objective of a successful calf and heifer rearing program is to achieve maximum first lactation milk production from an animal who produces her first calf at 24±2 months of age. The two graphs below illustrate standard growth curves for Canadian Holstein females between 1 and 24 months of age. As shown, growth of individual animals will vary, depending upon both genetics and nutrition.

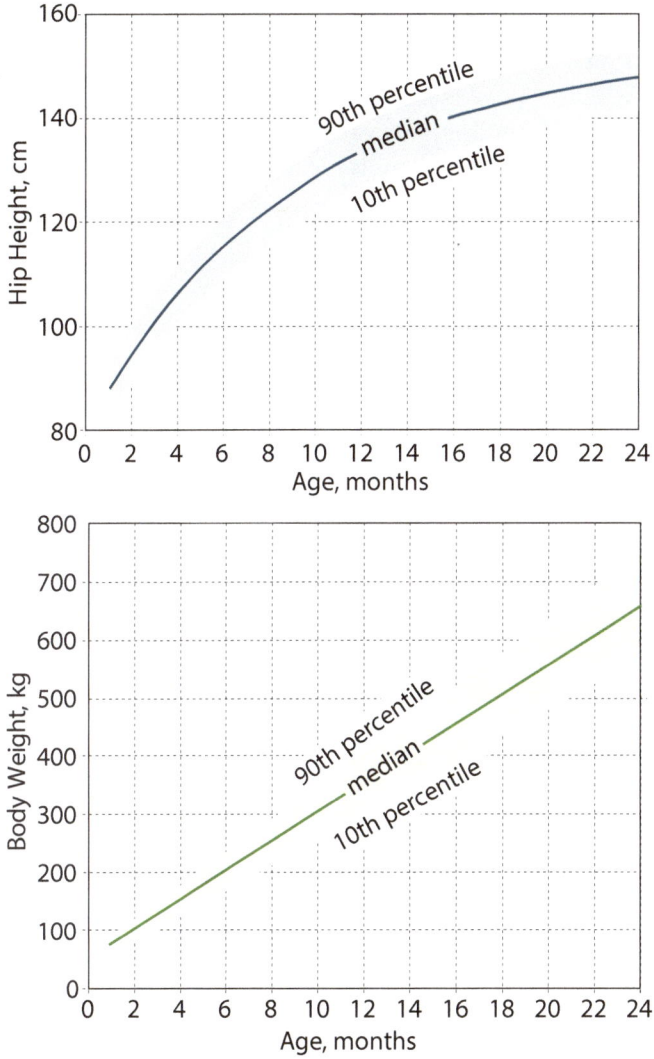

Figure 6.1: Suggested hip height (above) and weight (below) targets for Canadian Holsteins. source: Lactanet.

Growth targets can be determined with reference to mature body weight (BW) which is the weight of a cow after she delivers her third calf. Table 6.1 provides body weight and average daily gain targets for heifers between birth and first calving.

	Percent of Mature BW	Holstein	Jersey
Body weights	%	kg	kg
Mature	100	700	520
Birth	6	42	31
Weaning	12	84	62
Conception	55	385	286
First calving prepartum	91	638	426
First calving postpartum	82	574	474
Average daily gains	------------ kg/day ------------		
Prepuberty	0.13	0.90	0.67
Postpuberty	0.10	0.69	0.51
Postpuberty + pregnancy	0.13	0.92	0.69

Table 6.1: Target body weight and average daily gain targets for heifers recommended in NASEM Dairy 8 publication.

Monitoring heifer growth

The success of a heifer-rearing program can be evaluated by monitoring the height and weight of calves and heifers and comparing the results against the targets described above. Although most dairy producers, consultants, feed industry representatives, and veterinarians can recognize underconditioned or overconditioned animals, it is difficult to visually determine whether a heifer's height or weight is on target and typical for her age.

To estimate a dairy animal's body weight in the absence of a scale, use a weight tape, which is accurate to within 5 to 7 percent of the actual body weight. Make certain that the animal is standing with her head upright. Pull the weight tape snug, but not too tight, around the heart girth just behind the front legs and withers as illustrated in figure 6.2.

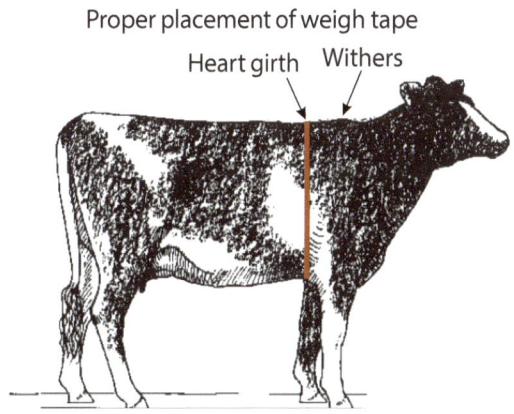

Figure 6.2: Proper placement of the weight tape for the determination of heifer weight. source: Penn State University.

When measuring a heifer's height, stand her on a clean, hard, level surface, hold her head upright and make sure that she is standing comfortably without pulling against her halter. Measure at the highest point of the hip as illustrated in figure 6.3, making sure that the horizontal part of the measuring stick is level.

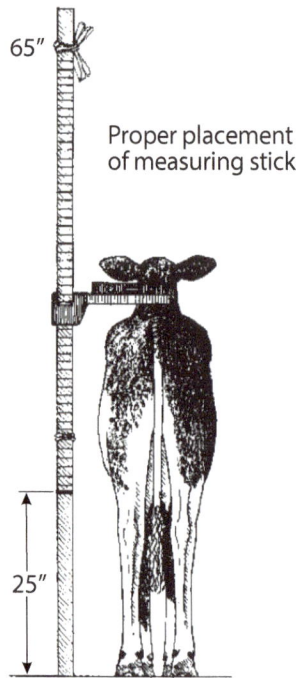

Figure 6.3 (right and below): Illustrating correct placement of the measuring stick when measuring heifer hip height. source: Penn State University.

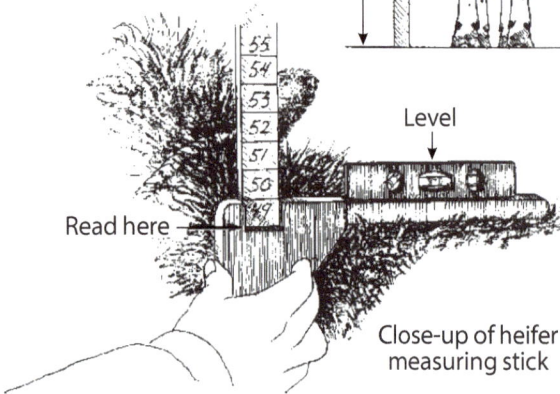

Feeding calves: birth to weaning

Colostrum

At birth, the calf has no protection against infectious diseases. Protection is acquired by consuming colostrum, the first milk produced by the dam and a rich source of both essential nutrients and antibodies (immunoglobulins: IgG) which provide resistance to

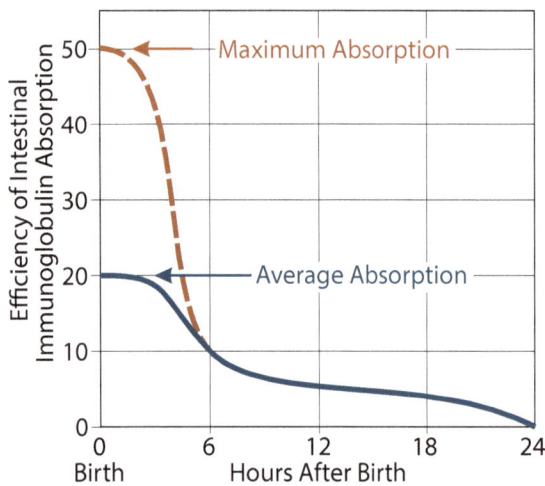

Figure 6.4: Intestinal absorption of colostral antibodies (immunoglobulins: IgG) declines rapidly after birth.

infection. To assure absorption of sufficient IgG, calves should receive 150 to 200 grams of IgG from colostrum or a colostrum replacement product within the first 24 hours of life. As illustrated in figure 6.4, the calf's ability to absorb antibodies declines rapidly after birth, varying significantly between calves. By 22 to 24 hours after birth, the calf is no longer able to absorb IgG into its bloodstream from its digestive tract.

Ideally, a 55 kg calf should be fed 3 litres of colostrum within an hour after birth and another 3 litres within the next 4 hours. Smaller calves can be fed proportionately less. Hand or esophageal feeding is the only way to assure adequate consumption.

Once colostrum is removed from the udder, subsequent milkings yield secretions which are much lower in nutrients and IgG than the first. This is shown in table 6.2. Since the concentration of IgG in colostrum is highly variable, routine evaluation of colostrum IgG content is recommended. Available tools are illustrated below.

Evaluation of colostrum quality

Good quality colostrum contains a minimum IgG concentration of 5% (50 mg/mL). In a survey of 13 Alberta dairy farms, IgG concentrations in 569 colostrum samples were measured using a colostrometer (right) and a Brix refractometer (below). The values determined by these methods were compared with values measured using a 'gold standard' lab test (RID). IgG concentrations varied from 8.3 to 128.6 mg/mL with 29% of samples having less than 50 mg/mL. The study also revealed that, although the colostrometer readings were better correlated with true IgG (RID) values, the user-friendly Brix refractometer is a more specific tool to detect colostrum of adequate quality and its measurement is not affected by colostrum temperature.

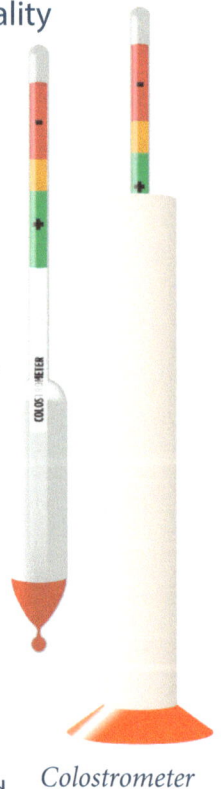

Colostrometer

Brix refractometer

| Item | Milking Number (cows milked twice daily) | | | | | |
| | 1 | 2 | 3 | 4 | 5 | 11 |
	Colostrum	---------- Transition Milk ----------				Whole Milk
Total solids, %	23.9	17.9	14.1	13.9	13.6	12.9
Total protein, %	14.0	8.4	5.1	4.2	4.1	4.0
Casein, %	4.8	4.3	3.8	3.2	2.9	2.5
Immunoglobulins, %	6.0	4.2	2.4	0.2	0.1	0.09
Fat, %	6.7	5.4	3.9	4.4	4.3	4.0
Lactose, %	2.7	3.9	4.4	4.6	4.7	4.9
Minerals, %	1.11	0.95	0.87	0.82	0.81	0.74
Specific gravity	1.056	1.040	1.035	1.033	1.033	1.032

Table 6.2: Average composition of mammary gland secretions from first 5½ days after calving.

The length of time between calving and first milking can have a significant effect on colostrum IgG content. In one study, IgG concentration was 27% lower in colostrum collected 10 hours after calving compared with colostrum collected 2 hours after. IgG concentration may also increase with maternal parity.

Needless to say, excellent hygiene is essential when collecting, storing and feeding colostrum. Its high nutrient content makes it an ideal medium for the growth of bacterial pathogens. Colostrum pasteurization at 60°C for 60 minutes has been shown to reduce bacterial counts while improving IgG absorption.

Feeding milk to calves

As described above, four to six litres of colostrum must be fed within the first few hours after birth for the calf to absorb the antibodies that will provide passive immunity for the first few months of life. After these initial feedings, colostrum and transition milk should be fed as long as they are available—usually for 3 to 4 days. Both contain significantly higher concentrations of nutrients (table 6.2) than the milk secreted later.

After receiving colostrum and transition milk for the first few days of life, calves are fed either milk or milk replacer until weaning. The majority of calves on western Canadian dairy farms receive waste (aka, 'hospital') milk—milk which is unsuitable for sale due to its poor quality or because the cow from which it is taken has been treated with antibiotics. Marketable (in-quota) milk may be fed when insufficient hospital milk is available but its high value compared to the cost of milk replacer makes it too expensive to feed routinely. There are risks inherent in feeding hospital milk:

• Solids content and nutrient composition can be quite variable, depending on the relative contributions of colostrum and transition milk (which increases solids, protein, and fat), dilution by water from washing procedures and milk from sick or treated cows.

• 'Hospital' milk may contain infectious organisms that can cause disease in the calves. *E. coli*, Johne's Disease, Mycoplasma, Bovine Virus Diarrhea and Pasteurella are of particular concern. Calf milk pasteurizers like the one shown in figure 6.5 reduce the risk of passing infection to the calves.

• Antimicrobial residues can inhibit the development of the bacteria that normally populate the calf's healthy gut. In addition, the development of antibiotic-resistant bacteria is a significant risk to both animal and human health. If possible, only pasteurized waste milk from cows not treated with antimicrobials should be fed.

When kept with her dam, a week-old calf will nurse about 10 times per day, consuming about 1 litre of milk each time—approximately 10 litres per day total. When allowed unlimited consumption of milk from an artificial teat, calves nurse an average of 25 times/day, consuming about 8.5 litres of milk per day at 7 days of age. However, traditional liquid feeding programs for calves deliberately restrict intake and growth. Many calves are fed as little as 2 litres of milk twice per day regardless of their size or the ambient temperature. The common rationale for this is that liquid feeds are expensive and that calves will compensate for limited liquid feed intake by increasing their intake of starter which will permit early weaning. Post-weaning compensatory ('catch-up') gain may occur as solid feed consumption increases rapidly. It is also believed that limiting liquid feed intake will decrease the risk of scours. There is ample evidence that, when a palatable, nutritious starter (see page 67) is offered beginning at 1 week of age, calves will rapidly start to increase its intake within a week and that its ultimate consumption will be inversely proportional

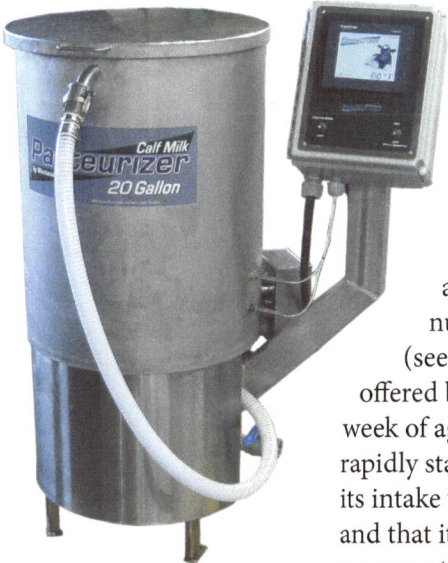

Figure 6.5: A calf milk pasteurizer.

to the amount of liquid feed offered. A number of research studies have demonstrated that faster preweaning growth rates resulting from higher combined intakes of liquid and solid feed are associated with higher first lactation milk yields.

It is strongly recommended that the traditional amount of 2 litres twice per day be increased to 4 litres twice a day for larger calves and when ambient temperatures below -10°C significantly increase maintenance energy requirements (table 6.3). This recommendation is in line with recommended best practices in the Canadian Code of Practice for the Care and Handling of Dairy Cattle (2009) which state:

- offer calves a minimum total daily intake of 20% of body weight in whole milk (or equivalent nutrient delivery via milk replacer) until 28 days of age (e.g., approximately eight litres per day for Holstein calves);
- increase milk intake when the environmental temperature drops below -10°C (increase all fluid diets by 25% in winter months).

Studies of the behaviour of calves offered varying amounts of milk or milk replacer revealed that calves offered less that 8 litres/day exhibited clear signs of hunger. Therefore, recommendation of this rate of feeding is also encouraged for the benefit of animal welfare.

Milk replacers

Although many dairy farms in western Canada rely on 'hospital' milk for feeding to their calves, milk replacers are fed in some situations:

- when the supply of hospital milk is limited;
- when it is recognized that hospital milk poses significant risk to the health of the calves;
- when the price of milk replacer is significantly less than the market value of marketable milk.

Milk replacers vary in composition and quality. 'All-milk' products contain only milk protein, usually derived from whey. When the price of skim milk powder is low, they may also contain milk caseins—the proteins that are used to make cheese. Less expensive milk replacers may contain soy-, wheat-, pea- or fish-

Ambient Temperature deg C	Birth to 3 Weeks	Older than 3 Weeks
	% increase in maintenance energy requirement	
45	38	40
40	28	30
35	19	20
30	9	10
25	0	0
20	0	0
15	9	0
10	19	0
5	28	9
0	38	18
-5	47	26
-10	56	35
-15	66	44
-20	75	53
-25	85	62
-30	94	71

Table 6.3: Effect of ambient temperature on maintenance energy requirements of young calves.

derived proteins, some of which are not well digested by young calves. Bovine or porcine plasma proteins may also be used. The digestibility of the protein in these sources is often negatively affected by the way they are processed. For example, spray drying produces less heat damage than either ring or drum drying.

Fat content is another important aspect of milk replacer quality, primarily affecting the energy value of the product. Typical sources include tallow, choice white grease, lard or palm oil. Emulsifiers, such as lecithin or monoglycerides may be included to improve mixing characteristics and fat digestibility. Vegetable oils and other lipids composed of high proportions of free fatty acids are poorly digested by calves. Both fat and protein quality affect the ease with which milk replacers mix with water and stay in suspension.

The primary carbohydrate in milk replacers is lactose from whey or skim milk powder. Other carbohydrates such as maltodextrin or starch are poorly digested, particularly in young calves, and are not recommended for calves less than 4 weeks of age.

In the past, antibiotics were commonly added to milk replacers in an attempt to protect calves from infection. This practice is no longer permitted in Canada except by veterinary prescription.

Milk replacers are typically fed at a daily rate of about 500 grams of powder dissolved in 4 litres of water, split into two equal feedings. It is recommended that this rate be increased by approximately 50% for larger calves and when the ambient temperature falls below -10°C. Increased feeding rates can be achieved either by adding more powder to the same amount of water or by feeding more mixed replacer at the usual dilution rate.

Nutrient requirements of calves fed milk or milk replacer only

Appendix table B1 contains recommended daily intakes of protein and energy for small (e.g., Jersey) and large (e.g., Holstein) calves consuming liquid diets only. Note that DMI in that table is not predicted dry matter intake but the DMI required to satisfy the recommended daily ME intake assuming the milk or milk replacer contains the defined ME concentration.

Calf starter

Dry feed intake drives rumen development (see page 46). The ruminal fermentation of starch yields high levels of volatile fatty acids which stimulate papillae development and physically effective fibre is essential to prevent the damaging effects of acidosis and parakeratosis (thickening of the papillae epithelium).

The sooner the calf is consuming dry feed, the sooner she can be weaned. Palatable, highly nutritious starter (figure 6.6) along with (but separated from) fresh water should be offered by a week of age. Small quantities, placed in a clean pail should be offered and cleaned out daily. Calves can be weaned when they are regularly consuming 1 kg/day of starter. Starter feeding should continue until the calf is consuming 1½ - 2 kg daily. Well managed calves in a clean, dry environment will normally consume this amount by about 8 weeks of age.

Table 6.4 provides typical nutrient specifications for a variety of calf starters. Calf starter should contain 18 - 25% crude protein (dry matter basis). Protein ingredients should be of high quality because, at this stage of development, the calf's digestive system is similar to that of a single-stomached animal—the rumen is only just starting to become functional. Soymeal is the most common choice of protein ingredients due to its good balance of amino acids and its palatability. Fresh ingredients, textured grains, added flavour and intact pellets with minimal fines will improve palatability and intake (figure 6.6). Calves fed starters containing a high proportion of fine particles will often benefit from being offered a source

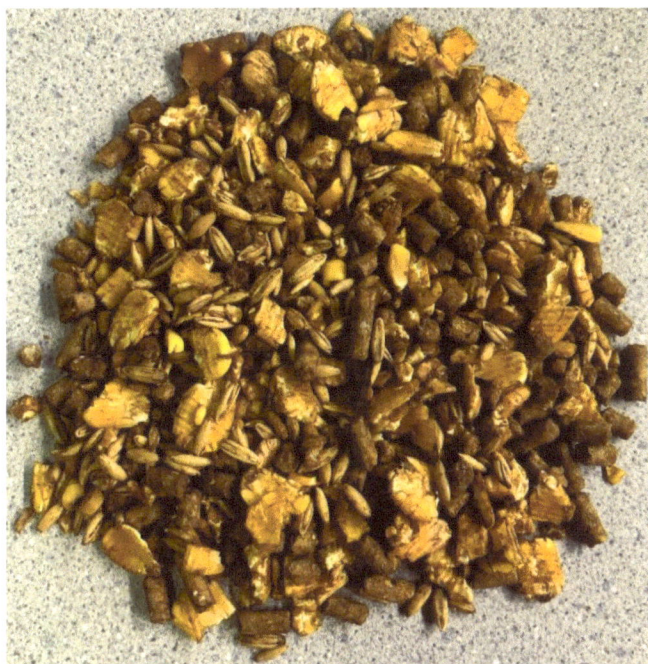

Figure 6.6: A typical 'textured' starter including steam-flaked grains, a pelleted supplement containing proteins, minerals and vitamins, a source of highly digestible fibre such as beet pulp or brewers grains and, usually, a small amount of molasses to improve palatability.

of chopped, high quality forage. However, this benefit may not be realized if the calves are bedded on straw or other edible organic bedding.

Calf starter should also contain a coccidiostat such as decoquinate or monensin to control coccidia, an intestinal parasite that frequently infects young calves.

Variables	16% CP Low Starch	18% CP Low Starch	18% CP High Starch	18% CP High Starch	22% CP Moderate Starch	22% CP High Starch
DM, % of As-fed	85.1	90.0	96.3	89.1	89.0	88.0
Starch, % of DM	15.1	20.7	36.9	39.0	25.5	32.9
CP, % of DM	18.8	20.0	18.7	20.2	24.7	25.0
ADF, % of DM	10.1	14.2	7.9	7.6	9.4	7.0
NDF, % of DM	24.8	29.5	18.9	15.9	16.3	13.7
IVdNDF48[a], % of NDF	49.1	60.0	53.1	55.4	65.3	59.6
Lignin, % of DM	2.91	2.19	2.09	1.83	1.61	1.61
Ash, % of DM	8.0	9.1	8.3	7.9	7.0	8.8
Starch, % of DM	15.1	20.7	36.9	39.0	25.5	32.9
WSC[b], % of DM	4.2	3.5	8.2	7.2	12.0	9.2
Crude fat, % of DM	6.9	5.1	3.9	3.6	3.3	3.2
DE base, Mcal/kg	2.67	3.21	3.22	3.19	3.62	3.38
ME, Mcal/kg	2.48	2.99	2.99	2.97	3.37	3.14

Table 6.4: Nutrient specifications for example calf starter feeds varying in crude protein and starch content. Note that the 'nominal' CP% (the CP% on the feed tag or label) is stated on the 'as-fed' basis, assuming 90% DM content.
[a]IVdNDF48: 48-hour in vitro NDF digestibility; [b]WSC: water-soluble carbohydrates

Group Housing and Calf Feeding Automation

Although it is still common practice to isolate newborn calves in individual hutches or pens, there is increasing interest in group housing for both newborns and older calves, for a number of reasons:

- Early research, reinforced by producer experience, led to the belief that housing calves individually would reduce the risk of pathogen transfer between calves. However, later studies have found little differences in health measures between calves housed individually and those housed in groups of 6-8 calves. Progress in hygiene management has undoubtedly contributed to the lower morbidity and mortality observed.
- It was (and perhaps still is) believed that housing and feeding calves individually would strengthen the bond betwee human caregivers and calves, a belief that has never been objectively confirmed.
- Public opinion is critical of individual housing for restricting free movement and social interaction among calves. Group housing facilitates both of these as well as other natural behaviours.
- Group-housed calves tend to consume more solid feed earlier, perhaps due to younger calves learning from their older pen mates.
- Group housing requires less feeding and management labour and lower capital costs than individual housing.

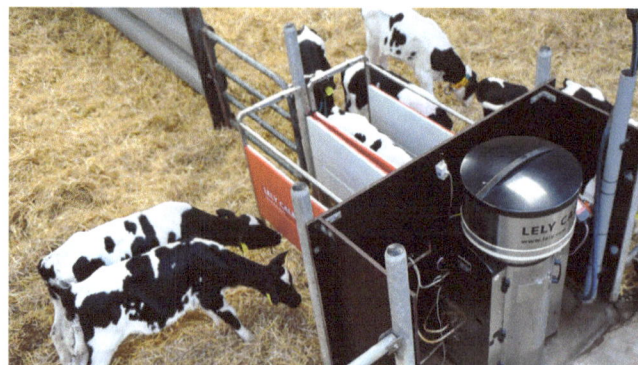

An automated liquid calf feeder.

One of the most significant advantages of group housing is that it creates the opportunity to automate feeding of both the liquid diet (milk or milk replacer) and starter. Today's automated liquid calf feeders facilitate:

- the feeding of milk or milk replacer with the ability to automatically mix replacer;
- control over the amounts and frequency of meals for individual calves;
- monitoring of individual intakes and feeding behaviours, raising alarms for behaviours that may provide an early indication of health issues.

Although automated feeding has the potential to reduce labour, excellent housing and hygiene management is no less important than with individually-housed calves.

Feeding hay to calves

In the past, it was believed that feeding hay to young calves was necessary to promote rumen development. The rationale was that physical 'scratch' was needed to start the rumen working. It is now well recognized that the main stimulus to rumen development is volatile fatty acid (VFA) production from feed fermentation. Because the amounts of VFA produced from grain are higher than those from forage, the rumen develops much faster when grain-based diets are fed, as shown in the photographs on page 46.

Another rationale for not feeding hay to calves before weaning is that the small amount they can consume contributes little to their nutrient requirements. Any hay that they do consume will limit starter intake and reduce total daily nutrient intake. Most of the hay they do not consume becomes bedding.

For these reasons, it has often been recommended that forage should not be offered to calves until after they are weaned.

Based on more recent research, current thinking recognizes that physically effective fibre is important even in preweaned calves to promote the growth of the muscular layer of the rumen and to maintain the health of the papillae—the small projections lining the inside of the rumen (shown in photos on pages 46 and 48). Rumen papillae can grow too rapidly in response to high levels of VFA. When this happens, they may clump together, reducing their absorptive capacity. Also, some 'scratch' is needed to prevent the formation of a surface layer of keratin (referred to as parakeratosis), which can also inhibit VFA absorption.

For calves offered pelleted starters and not bedded with straw, the current recommendation is to offer chopped, high quality forage (preferably alfalfa) separately from the starter or mixed with the starter at no more than 5% of total dry matter intake. Calves offered textured starter and bedded on straw or other organic bedding will not likely benefit from the availability of additional 'structured fibre'.

Weaning

Many producers wean calves when they reach a particular age—often 6-10 weeks. Although this might be effective for most, some calves are not ready to be weaned at that particular age while many are ready much earlier. The critical question is: "How well developed are their rumens?". Since you can't actually monitor rumen development, the next best indicator is dry feed intake.

A healthy, small breed (e.g., Jersey) calf routinely consuming 1.25 kg of starter daily is ready to be weaned; large breed (e.g., Holstein) calves should be consuming 1.5 kg/day. The age at which she has achieved that level of intake is mainly a function of management. Calves raised in a clean, dry environment who are offered fresh, nutritious and palatable starter and water in clean containers each day from their first week of life can often consume those levels of starter by as early as 4 weeks of age and certainly by 6 weeks. Gradual weaning over 4 - 10 days will minimize the risk of growth setbacks that can result from weaning too abruptly.

Nutrient requirements of calves fed milk or milk replacer and starter

Appendix tables B2 and B3 provide examples of recommended energy and protein requirements for small- and large-breed calves fed milk or milk replacer and starter at two different ratios. Note that in those tables, DMI is dry matter intake assuming a mean ME density needed to meet ME requirement; it is not a prediction of actual DMI. Appendix table B4 provides recommended levels of minerals and fat-soluble vitamins in milk replacer, starter and grower; table B5 gives recommended concentrations of water-soluble vitamins in milk replacers. These recommendations are derived from the NASEM Dairy 8 computer model described in Appendix C.

Feeding heifers from weaning to breeding

Between weaning and conception at 13 - 15 months of age, the heifer will gain from 200 to 400 kg and grow in height by 30 to 50 cm. Target growth rates recommended in the NASEM Dairy 8 publication are presented in table 6.1.

Starter should continue to be offered immediately after weaning along with high quiality hay. If calves are not weaned until 8 to 10 weeks of age, a limited amount (500 grams/day) of the best hay available should be offered from about 6 weeks of age. A grower ration, containing 16 - 18% crude protein (dry matter basis) can replace starter when calves are around 4 months of age and consuming 2 kg of starter daily. The grower protein level and the amount to feed will depend on hay consumption and quality. Recognize the difference between the amount of hay offered and the amount consumed—much of the hay offered can become bedding if feeders are not well designed.

Beginning at about 6 months of age, heifers can be fed a silage-based total mixed ration (TMR), where all ingredients are combined in a single mix. This strategy provides better control of diet composition and intake than does separate feeding of forage and concentrate.

The few months before a heifer reaches puberty (normally at 9 - 10 months of age) is a critical time for offering a well-balanced ration, designed to support lean tissue growth. If the diet contains excess energy, the heifer will begin to gain body condition in the form of fat (adipose tissue). Fat that is laid down in the developing mammary gland can have negative effects on milk secretory tissue development and milk production later in life (see sidebar below).

Nutrient requirements of heifers from weaning to breeding

Appendix table B6 provides energy and protein requirements of heifers fed only solid feeds between weaning and breeding.

Nutrition Affects Mammary Development

Before first conception, development of the heifer's mammary gland consists of expansion of the milk-producing tissue and duct network (the parenchyma; PAR) within the mammary fat pad (MFP). The effects of nutrition on the relative growth of these two components of the gland can affect future milk production potential. Nutrition is particularly important at 2 stages of prepubertal heifer development:

- A high plane of nutrition pre-weaning supports the development of both PAR and MFP which, combined, have the potential to grow at 50 times the rate of other tissues between 1 and 3 months of age. By 5 months of age that rate has fallen significantly but continues to remain higher than the rate of body weight gain until puberty. This supports the argument for maximizing liquid and starter intakes in the first few months of life.
- In contrast, a high plane of nutrition pre-puberty can negatively affect PAR development in two different ways. First, a high energy diet in this period can increase the deposition of fat in the MFP without affecting the age-dependent rate of PAR development. It is thought that the higher fat content of the mammary gland can impair first lactation milk production. Second, a rapid increase in body weight at this time can result in earlier puberty, at which time PAR development slows to a rate more in line with that of other tissues. This may ultimately lead to a reduced proportion of milk-secreting tissue (PAR) in the mammary gland, negatively affecting later production potential.

Feeding bred heifers

Although first calving at 22 to 24 months is considered the best compromise between the cost of raising a heifer and her lifetime income potential, in 2021 only the top 25% of DHI herds in western Canada achieved that goal. If a heifer is to calf before 24 months, she must be bred by 15 months of age. By this age she should weigh 55% of her predicted mature body weight (MBW).

During a 278 day pregnancy, the Holstein heifer's own body will grow by 190 to 210 kg, an average daily gain of around 0.7 kg, assuming her potential is above the 50th percentile suggested by standard growth curves for Holsteins. Targets for smaller breeds will be slightly lower. For the first two-thirds of pregnancy (up to 185 days), nutrition is required to support the heifer's own body growth only. If high quality forage is fed, little more than supplemental salt, minerals and vitamins, along with fresh water, are required. Lower quality forages may need to be supplemented with concentrates providing additional energy and protein.

During the last third of pregnancy, the fetus and associated tissues are growing rapidly, as illustrated in figure 6.7. Nutrient requirements increase significantly and, if total mixed rations (TMRs) are fed to the lactating herd, this is a good time to introduce a TMR to the bred heifers. Otherwise, a concentrate mix can be fed separately to supplement forages fed *ad libitum*.

At approximately 3 weeks before their expected calving dates, pregnant heifers are often moved to a pen housing close-up dry cows. In a small herd, it may be considered efficient to manage these 2 groups together. However, this practice often puts the heifers at a disadvantage to the mature, more dominant older animals who may prevent heifers from consuming the amount of feed they require. Insufficient feed

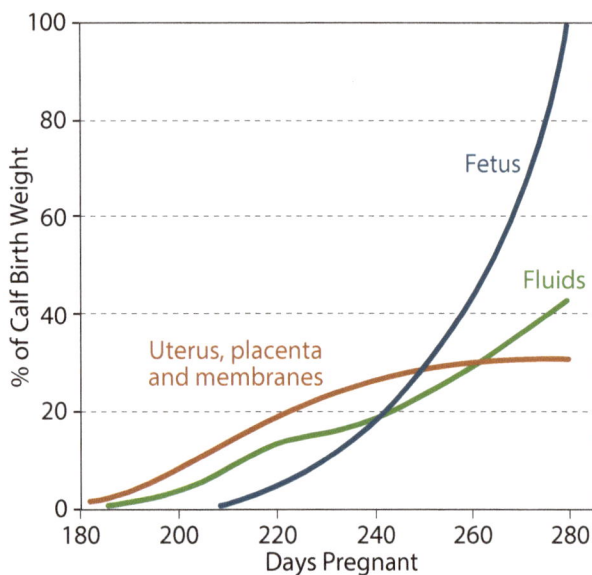

Figure 6.7: Conceptus growth patterns.

consumption at a time when nutrient requirements are rising can increase their risk of metabolic disorders and prevent heifers from reaching their first lactation production potential. Ideally, heifers should be managed separately in both the close-up period and in first lactation.

Nutrient requirements of weaned calves and growing and bred heifers

Appendix tables B7 and B10 provide examples of nutrient concentrations required in dietary dry matter needed to meet the requirements for growing large-small-breed heifers, respectively, up to first calving. These recommendations are derived from the NASEM Dairy 8 computer model described in Appendix C.

Lactating and dry cows
Feeding lactating cows

The nutrient requirements of lactating cows vary according to parity (lactation number), cow size, body condition, stage of lactation and production level. With this in mind, it might appear most efficient to feed animals individually, in an attempt to satisfy each cow's particular requirements. In the past, many herds were fed with this objective in mind, particularly those housed in tie-stall and stanchion barns. Computer grain feeding stations also made individual concentrate feeding possible in free-stall barns. In this system, each cow wears an individual electronic identification tag and daily allowances for each are programmed into the computer. However, one of the shortcomings of both of these feeding strategies is that there is seldom any control over the forage intakes of individual cows. Therefore, total daily nutrient intakes are difficult to determine. Many cases of ruminal acidosis in these systems can be attributed to the consumption of too little physically effective fibre.

Few herds in western Canada are housed in tie-stall or stanchion barns and virtually all free-stall computer grain feeding systems have now been abandoned in favour of feeding total mixed rations (TMRs).

Total mixed rations

Many producers were initially reluctant to accept TMR feeding as it seemed contradictory to the premise of widely varying requirements among individual cows. The ability of a single ration to satisfy the nutrient requirements of both high- and low-producing cows of all sizes and at all stages of lactation was initially questioned. However, it is now realized that the feed intake of each cow is a primary determinant of nutrient intake, varying in proportion to cow size, stage of lactation and production potential. Since the late 1990s,

TMR feeding has become the norm for virtually all free-stall housed herds as well as many of those few still housed in tie-stalls and stanchions.

One of the main advantages of TMR feeding is its potential to ensure that all cows are consuming the prescribed balance of forage and concentrate as well as adequate fibre. However, this potential is often not realized because of feed sorting. Some cows are capable of separating fine, dense particles out of a dry TMR by repeatedly 'nosing' the feed from side to side. Longer forage stems are nosed aside while small, nutrient-dense concentrate particles that fall to the floor are readily consumed. It is therefore important to ensure that forages are adequately but not excessively chopped (see sidebar above), the TMR is well mixed and that all ingredients are wet enough to prevent such sorting. Total TMR moisture level should be in the 45 - 65% range. If dry hay is included in the ration, mixing several hours before feeding will allow the hay to absorb moisture from the other ingredients, reducing the likelihood of separation. If the moisture level is lower than 45%, additional water or up to 4% liquid molasses can be thoroughly blended into the mix, again allowing several hours for it to absorb before feeding.

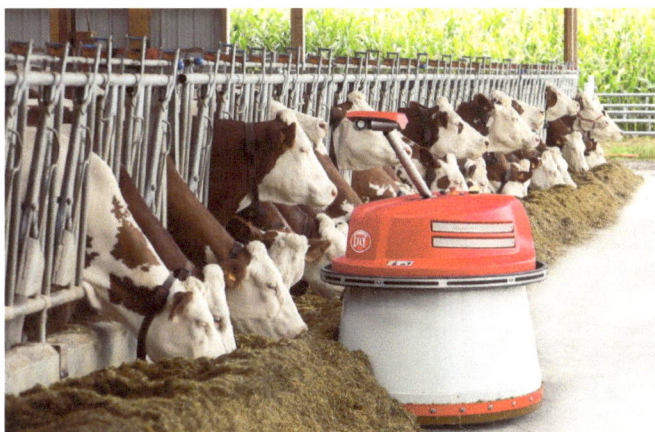

Figure 6.8: A robotic feed pusher.

In most drive-through free-stall facilities without enclosed feedbunks, cows will push feed away from the feed barrier beyond their reach. In an attempt to increase feed consumption, many producers find 'pushing up' to be a useful strategy. Pushing up not only places more feed within their reach, but the activity often provokes the cows' natural curiosity and attracts them to the bunk. Figure 6.8 illustrates a robotic feed pusher designed specifically for this purpose.

Most free-stall housed herds in western Canada offer a single total mixed ration (TMR) to all lactating cows. A few larger herds will mix 2 TMRs—a high ration and a low ration. The high ration is typically fed to a group including cows from immediately after calving until some time near mid-lactation. Cows past mid-lactation are put into a group offered the low ration. In some cases, fresh cows are put into the low group for 1 - 2 weeks to ease the transition to the high ration. In other herds, just-fresh cows are offered a third TMR for 10 - 20 days, similar to that fed to the close-up dry group. Each TMR is formulated to satisfy the nutrient requirements for the target average production level of cows in that group.

The disadvantage of multiple TMRs in small herds is that, due to negative social interactions, feed intake and production often fall when cows are moved from one group to another. This is most common when small numbers of animals are moved into larger groups.

Partial mixed rations and automated milking

An increasing number of western Canadian dairy herds are adopting an automated milking system (AMS) where cows are offered a portion of their daily ration in a robotic milker. The remainder, referred to as a 'partial mixed ration' (PMR) is typically offered *ad libitum* in a feed bunk or drive-through feed alley.

Three basic system designs influence cow traffic flow through the robot:

- *Free flow* systems allow cows to move freely between the free stall, feeding and milking areas.
- *Milk-first guided flow* systems require cows exiting the free stall area to pass through a selection gate. Those eligible to be milked are guided to the robot. Otherwise they are guided to the feeding area and can only re-enter the resting area through a one-way gate.
- *Feed-first guided flow* systems require cows exiting the free stall area to pass through a one-way gate to the feeding area. A selection gate controls flow from the feeding area to the robot or back to the free stall area.

Several other variations on these guided flow systems are also possible.

Based on research conducted at the University of Saskatchewan's Rayner dairy and their extensive review of feeding trials in AMS systems elsewhere, the following recommendations can be made:

- The portion of the daily ration offered in the robot should be pelleted to prevent separation of ingredients.
- The total daily amount offered in the robot should not exceed 4 - 5 kg of dry matter and robot feeders should be routinely checked for orts (unconsumed feed). If significant orts are routinely found, the amount offered should be decreased.
- The complete diet (PMR + robot-fed) should be formulated as if it were a TMR with the robot portion as a discrete ingredient.
- Increasing the amount of ration offered in the robot in an attempt to supplement the amount of PMR consumed is not effective. Higher amounts consumed in the robot will generally reduce the amount of PMR consumed.
- There is no data to suggest that ration allocation in the robot should be dependent on whether the cows are housed in a free-flow or a guided-flow system.

Trials to evaluate the effect of ingredient composition on consumption of robot-delivered feed have been inconclusive. A palatable pellet formulated to complement the nutrient composition of the PMR is recommended.

Lactation curves

When discussing lactation curves, we often present smooth graphs like those illustrated in figure 6.10. Most producers will realize that daily production records for individual cows look nothing like these curves. The production from individual cows fluctuates up and down from one milking to the next. Figure 6.9 shows typical fluctuations in AM, PM and daily (24 hr) milk yields. Superimposed on the daily record are two possible sequences of DHI test days.

The 'typical' lactation curves shown in figure 6.10 represent average yields by days in milk calculated from several thousand DHI 305-day lactation records, fitted to 'best fit' mathematical functions for each parity (lactation number). After calving, production rises rapidly, reaching a peak between about a month and 3 months in milk, depending on parity and production level. Production declines steadily from peak until dry-

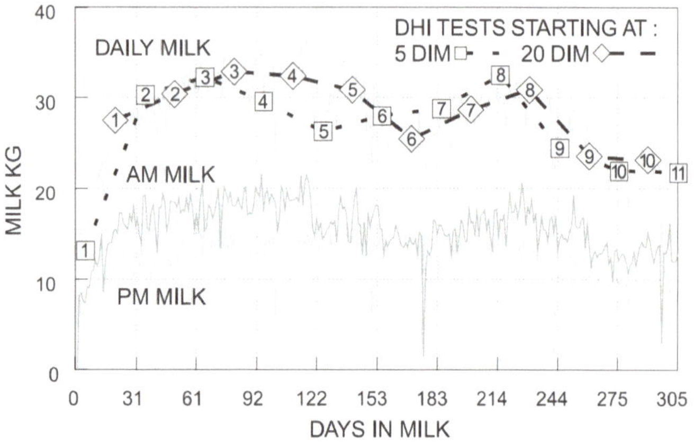

Figure 6.9: Raw daily milk production record for an individual cow illustrating variation from milking to milking and from day to day. Lactation curves computed from 2 different sequences of DHI test days are superimposed on the daily (24-hr milk) yield data.

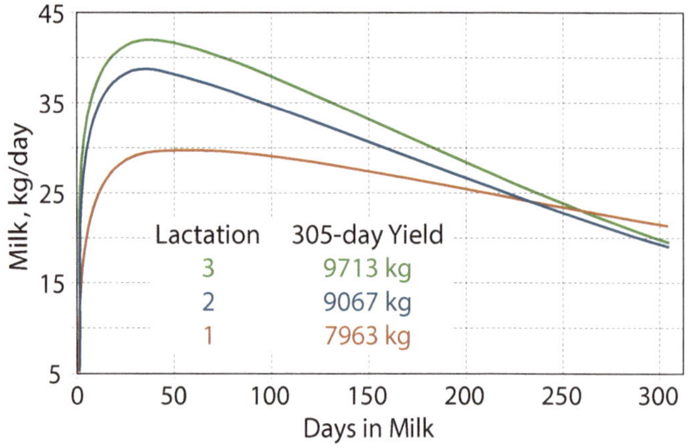

Figure 6.10: The shape of the lactation curve varies in first, second and third lactations.

	305-Day Production Level, kg							
	Over All	5000 -5999	6000 -6999	7000 -7999	8000 -8999	9000 -9999	10000 -10999	11000 -11999
Peak Milk Kg :								
Lactation 1	29.6	23.4	26.0	28.7	31.7	34.8	37.7	41.0
Lactation 2	38.7	29.2	32.2	34.9	37.5	40.5	43.5	47.0
Lactations 3+	42.0	30.1	33.4	36.2	39.2	42.0	45.1	48.0
Days in Milk at Peak :								
Lactation 1	56	39	44	48	57	64	69	76
Lactation 2	35	27	27	28	32	35	38	41
Lactations 3+	37	27	31	31	32	36	38	41
Persistency (66-305 DIM) :								
Lactation 1	96.5	95.5	95.8	96.1	96.4	96.7	97.0	97.1
Lactation 2	93.8	92.4	92.7	92.9	93.4	93.6	93.9	94.1
Lactations 3+	93.5	91.8	91.9	92.3	92.7	93.1	93.3	93.7

Table 6.5: Typical lactation curve characteristics for western Canadian Holsteins.

off. The rate of this decline, referred to as persistency, is defined as milk yield on a particular day expressed as a percentage of milk yield one month (30.5 days) earlier. Typical peak yields, days in milk at peak and average 66 - 305 day persistencies for western Canadian Holsteins are given in Table 6.5.

Social stress from regrouping of cows, feed changes, weather events and health status, including lameness, can all lead to marked changes in the shapes of lactation curves.

Negative energy balance

The relationship between dry matter intake (DMI) and milk production can be described as follows:
• Genetic milk production potential drives DMI but changes in DMI lag behind changes in potential production throughout the lactation cycle.
• The actual production level achieved depends on nutrition, environment and other factors controlled by management. Ideally, the cow will achieve her genetic potential but, under less than ideal conditions, this potential will not be realized.

As illustrated in figure 6.11, milk output rises more rapidly than DMI in early lactation. In most cases, the energy expended by the cow exceeds her energy intake. The result of this negative energy balance is that she draws energy from her body fat stores, losing weight and body condition, an indication of the amount of body fat she carries (see page 74). Later in lactation, positive energy balance allows the recovery of body weight and condition.

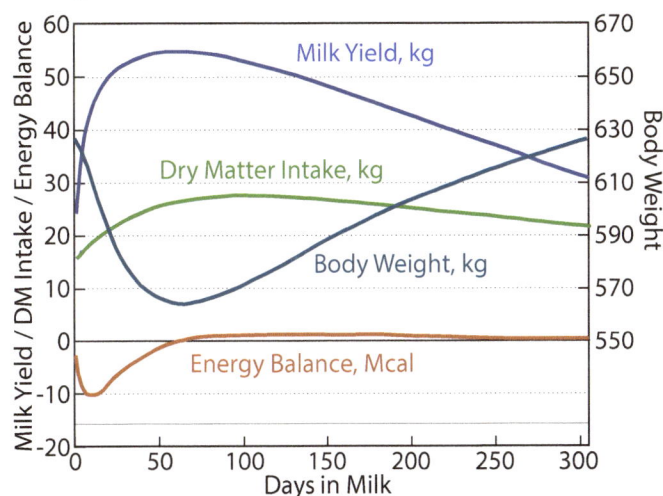

Figure 6.11: Relationships between changing milk yield, body weight, dry matter intake and energy balance as lactation progresses.

Dietary effects on milk composition

Canada's dairy supply management system attempts to match the supply of milk components produced on the farm with consumer demand for those components.

However, there are limitations to the producer's ability to influence the composition of the milk produced. The butterfat concentration in bulk tank milk can range from as low as 2.5 kg/hL to over 4.6 kg/hL, depending on breed, season, diet and herd average days in milk. But the protein and lactose concentrations are much less variable (8.1 - 10.3 kg/hL combined). Over the past several decades, there has been greater consumer demand for dairy products containing higher levels of butterfat (BF) than for those containing protein and lactose ('solids non-fat': SNF). This has resulted in a 'structural surplus' of SNF relative to butterfat (BF) and an accumulation of skim milk powder stocks. In an attempt to bring the production of SNF and BF into closer alignment with consumer demand, the industry has implemented mandatory maximum SNF:BF ratios. In Alberta, for example, the 2022 ratio was 2.4167 meaning that a producer with a bulk tank milk SNF concentration of 9 kg/hL had to maintain a BF concentration of at least 3.72 kg/hL.

Available dietary options for increasing milk BF concentration include:
• *increasing the proportion of fibre in the diet.* Rumen microbes that digest fibre produce higher proportions of acetic acid than those that digest starch or protein. Acetic acid is an important precursor for the synthesis of fatty acids in the mammary gland. The down side of this strategy is that fibre (particularly fibre in mature forages) is more slowly digested that other carbohydrates, resulting in reduced feed intake and lower milk yield.
• *supplementing the diet with saturated long chain (≥C16) fatty acids (FAs), particularly palmitic acid (C16:0).* While the mammary gland synthesizes most milk fat FAs with chain lengths of 16 carbons and less, longer chain FAs in milk are derived mainly from the diet and from mobilization of body fat stores. Although palmitic acid is the predominant milk FA (30 - 40% of total FA), the addition of 1.2 - 2.0% enriched (>85%) C16:0 supplement can increase milk C16:0 content by 5.5 - 12.3% with a corresponding increase in milk fat content of 0.2 - 0.5 kg/hL. Part of this increase in milk fat concentration may result from the consistently-observed increase in NDF digestibility with C16:0 supplementation.
• *diet supplementation with rumen-protected methionine (RP-Met).* A recent review concluded that supplementation of diets with 7.5 - 12.5 g/day of RP-Met could increase milk fat concentration by 0.4 - 1.7 kg/hL. However, that level of supplementation also increased milk protein concentration by 0.4 - 1.4 kg/hL rendering this strategy of limited value in reducing the herd's SNF:BF ratio.

Condition score targets

Condition scoring is a simple, subjective method of assessing the amount of body fat an animal is carrying. A score of 1 is assigned to an emaciated cow; 5 to a very fat animal. Each score increment is associated with an approximately 9.4% change in empty body weight. Figure 6.12 details the appearance of 5 key areas on the cow which are used to determine the score.

The main objective of condition scoring is to have cows calving with a score of 3¼ to 3¾. In this condition they are carrying adequate reserves which, combined with maximum nutrient intake in early lactation, allow them to reach their highest possible peaks. If normal persistencies can be maintained in mid and late lactation, every extra kg of peak yield can result in an additional 200 - 250 kg of production over the lactation. Condition score targets as the cow progresses through her lactation cycle are shown in figure 6.13.

Excessively fat cows (score >4), have been shown to consume less feed before and after calving, which negates the potential contribution of their extra body reserves. These cows also have a higher incidence of metabolic problems including left displaced abomasum and ketosis after calving.

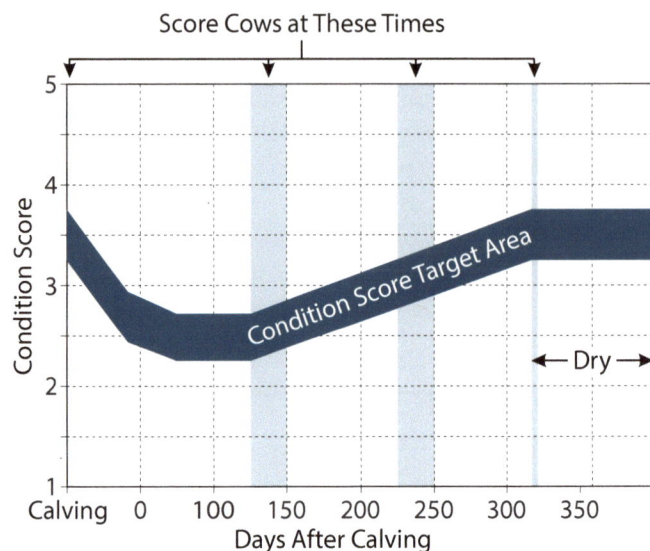

Figure 6.13: Condition score targets.

Feeding dry cows

Sixty days is the recommended optimum dry period length. During the first 5 weeks, the mammary gland will involute after the previous lactation and the rumen is rejuvenated by feeding a high-forage diet. Cows should enter the dry period with a condition score of 3¼ to 3¾ and that score should be maintained through to calving.

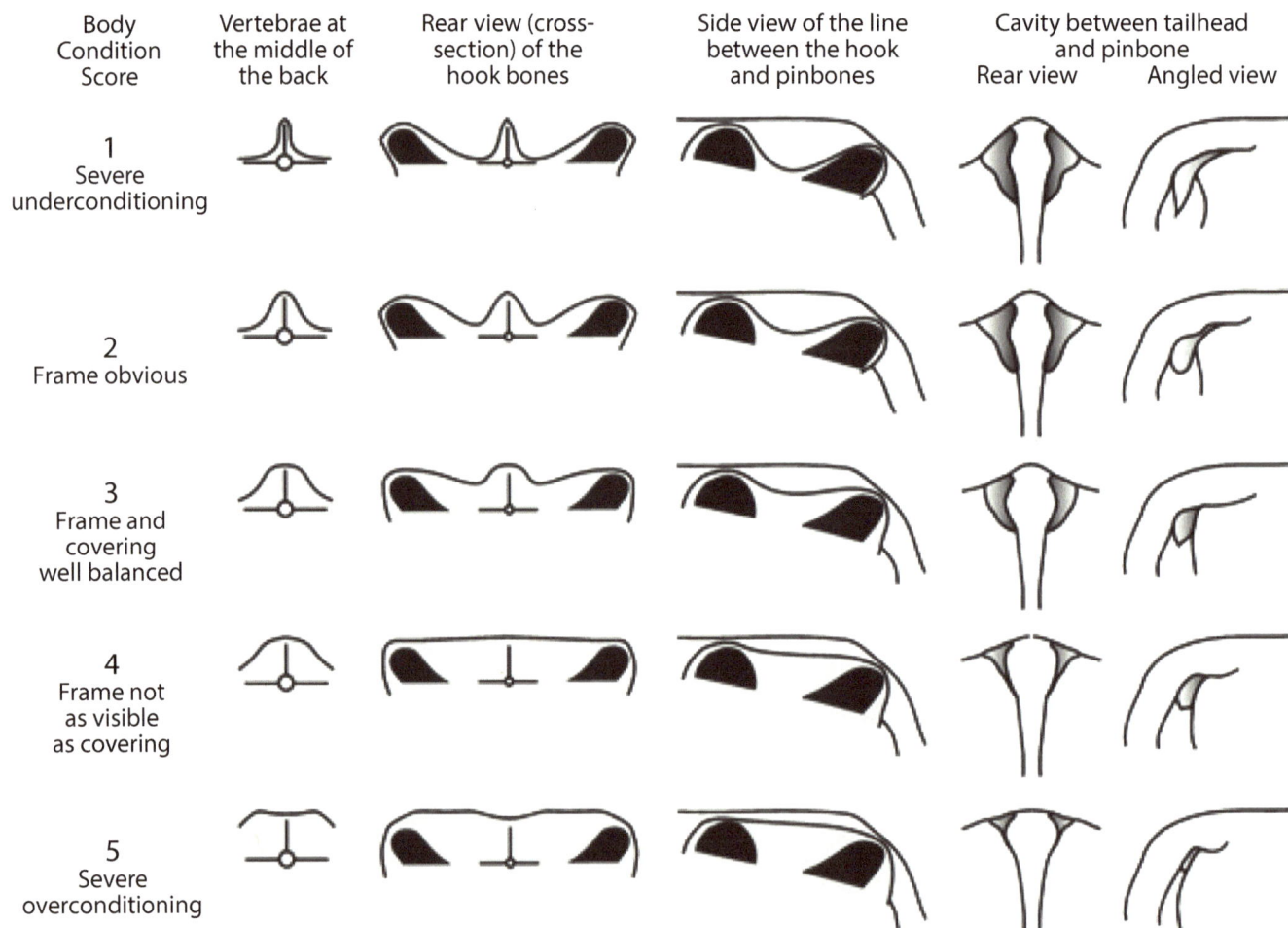

Figure 6.12: Schematic of fat presence in 5 areas examined to assess condition score.

The Goldilocks Strategy: not too little, not too much, just enough nutrition

The basic principle behind the design of the 'Goldilocks' diet is to provide the nutrients required by the dry cow in a ration that allows the cow to satisfy her fill capacity. Here is how that principle is applied in practice:

- The ration is initially formulated based on the dry matter intake (DMI) predicted by the equations described in chapter 7, pages 81 - 82.
- A maximum amount of low quality forage (preferably straw) is used to increase ration bulk such that energy intake is limited by the cow's rumen fill capacity.
- Wheat straw is preferable to either barley straw, oat straw or other forage due to its coarse, brittle stems which result in more uniform particle sizes when chopped for incorporation in the TMR.
- The forage should be pre-chopped to yield particles that are 5 cm or less in length before mixing into the TMR. Few TMR mixers are capable of producing particles of this size when large proportions of straw are included without over-processing other ingredients.

- Depending on the amount of straw included, the TMR moisture level may need to be increased by adding water and allowing the mix to stand for several hours to promote uniform moisture distribution. The strategy will fail if cows are able to sort out the straw from the TMR.
- Although appendix tables B9 and B12 suggest different nutrient requirements for far-off cows (60 - 21 days pre-calving) and close-up cows (less than 21 days pre-calving), there are disadvantages to managing 2 dry cow groups, particularly for smaller herds. Several research trials and some producer experience has demonstrated that a single, full dry period compromise diet incorporating the Goldilocks strategy is possible.
- Because of the bulkiness of these diets, cows will spend more time eating. Therefore, it is important that their access to feed is not limited in any way.

The fetus and other products of conception grow rapidly during the dry period (figure 6.7) and the mammary gland expands in preparation for lactation. Although the nutrient requirements for conceptus growth are minor compared to those for milk production, it is important to recognize that the dry cow has requirements beyond those required for maintenance.

During the last 3 weeks before calving (the close-up period) the cow is prepared for the next lactation. Traditionally, dietary nutrient concentrations have been increased for several reasons:

- to satisfy the increasing requirements of conceptus growth and mammary development;
- to compensate for a decrease in dry matter intake (DMI) which often occurs in the week before calving;
- to adapt rumen microbes to the higher level of dietary starch they will see in early lactation.

Declining pre-partum DMI has been thought to be the trigger for a number of metabolic disorders that occur around calving, including ketosis, fatty liver syndrome, milk fever, retained placenta and displaced abomasum. Although its cause is unclear, hormonal changes and the stress caused by changing diets, uncomfortable housing and movement into unfamiliar environments and social groups are thought to contribute. In addition, prefresh cows with body condition scores greater than 4 have been found to consume up to 8% less dry matter than their lower scoring herdmates.

Over the past 20+ years, thinking about the ideal diets for dry cows has turned to providing adequate but not excessive nutrient intakes by including poor-quality forage in the form of straw to allow cows to consume to their physical fill capacity. The implementation of such so-called 'Goldilocks' diets is described in the sidebar above. Implementation of this feeding strategy has been reported to significantly reduce metabolic disorders in the transition into lactation.

Dietary cation-anion difference (DCAD)

The principle behind DCAD and its application to dry cow diets are described on page 61. Although it has been successfully applied in a number of research trials, the results of its application on-farm have been highly variable. One of the probable reasons for this is that, to properly implement DCAD, it is necessary to have accurate measures of the relevant cations and anions in all dietary ingredients—'book' values are not appropriate.

For the majority of producers in western Canada, milk fever prevention is achieved by routinely administering a calcium supplement to fresh cows.

Nutrient requirements of lactating and dry cows

Appendix tables B8 - 9 and B11 - 12 present example nutrient requirement recommendations for lactating and dry Holstein and Jersey cows, respectively. These recommendations are derived from the NASEM Dairy 8 computer model described in Appendix C.

Effects of nutrition on health

The maintenance of good health requires adequate intake of a well-balanced diet. Here is a short summary of the roles played by specific nutrients in common health problems:

Energy: When energy intake is lower than energy demand (negative energy balance, see page 73) the animal mobilizes body fat. When this occurs in late pregnancy or early lactation, mobilized fat may accumulate in the liver and result in the production of ketone bodies (acetone, acetoacetic acid and β-hydroxybutyrate), leading to ketosis. Excess energy in the form of rapidly-fermentable carbohydrates (e.g., starch) can lead to ruminal acidosis, the result of high levels of acid produced during rumen fermentation.

Protein: Adequate degradable protein is required to support an active population of rumen microbes (bacteria, protozoa and fungi). These microbes are responsible for digesting the carbohydrates which provide most of the animal's energy. The combination of dietary and microbial protein digested in the small intestine must satisfy the animal's requirements for maintenance, production and the synthesis of new proteins needed to fight infection. Excess protein places an extra load on the liver where surplus nitrogen must be converted to urea before being excreted.

Fibre: Physically effective fibre (structured roughage) requires the ruminant to chew and salivate, releasing buffers which help to prevent ruminal acidosis. Rumen microbes digest fibre more slowly than starch and other non-fibre carbohydrates, resulting in lower and more stable acid levels and less risk of acidosis. Excess fibre limits intake, which may also limit the availability of other nutrients.

Calcium: At calving, calcium requirements increase dramatically due to the new demand for milk synthesis. Initially, most of this increased requirement is derived from bone. When the mechanism involved in mobilizing bone calcium fails to 'turn on' properly, blood calcium concentration may fall below the level required to maintain muscle tone (hypocalcemia), causing the cow to go down with milk fever.

Magnesium: Grass tetany is a result of a low blood magnesium concentration (hypomagnesemia), caused by low magnesium or high potassium intakes.

Trace minerals: Copper and zinc are important in the maintenance of hoof health, both because they are required in the synthesis of hoof tissue protein (keratin) and because they act as antioxidants in support of the immune response to infection. Other minerals important to immune function include selenium, manganese, and iron.

Vitamins: The role of vitamin A in the immune system is to maintain the integrity of the epithelial (surface) tissues of the body which block the entrance of pathogens. Vitamin E is essential in increasing antibody production in response to the presence of a pathogen, and vitamin C is believed to benefit immune function possibly via improved neutrophil (a type of white blood cell) function, however this is typically only seen in non-ruminant calves.

Ruminal acidosis

Ruminal acidosis occurs when excess acid is produced by rumen fermentation, causing rumen pH to fall below 5.6. Subacute ruminal acidosis (SARA), where pH is between 5.2 and 5.6, is probably the most common metabolic disorder in dairy cattle, affecting about 20% of cows in the first 120 days in milk. SARA causes mild diarrhea, reduced dry matter intake and hemorrhages in areas of the hoof where new tissue is formed (laminae), leading to laminitis. Fibre digestion is reduced when rumen pH falls below about 5.8 as the environment becomes less favourable for the survival of fibre-digesting microorganisms. Acute acidosis, where rumen pH falls below 5.2 is less common but much more serious. Affected animals are depressed, off-feed, have an elevated heart rate, diarrhea and may die.

Acidosis can result from excess feeding of grain or other non-fibre carbohydrates (NFC), a rapid increase in the dietary content of NFC or insufficient rumen buffering due to inadequate chewing and salivation. When cows are introduced to lactation diets after calving, the higher concentrate and lower physically effective fibre content of these diets results in a higher rate of acid production. This, combined with reduced absorptive capacity of the rumen coming out of the dry period, can also cause rumen pH to drop.

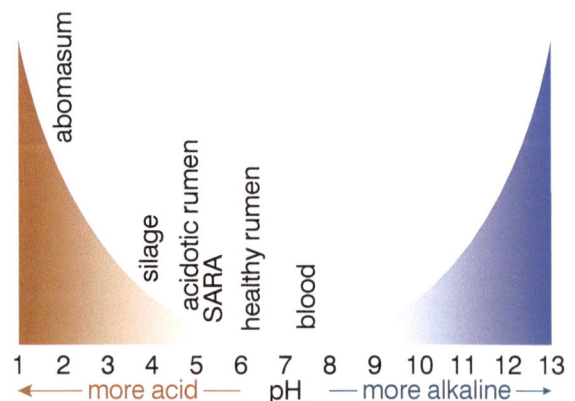

Figure 6.14: pH is a measure of acidity and alkalinity. Pure distilled water has a pH of 7—it is neither acid nor alkaline. Each unit of pH below 7 represents a 10x increase in acidity; each unit above, a 10x increase in alkalinity. Ideal rumen pH is in the 5.8 - 6.4 range.

Bloat

The obvious sign of bloat is enlargement of the left side of the animal, immediately behind the last rib. Bloat results when gases produced from rumen fermentation are unable to escape through the esophagus. Three types of bloat can occur:

Figure 6.15: Bloat.

- pasture (frothy) bloat—seen in the early spring and summer due to pasturing on young, lush, usually legume, forages (e.g., alfalfa); pectin in the forage will react with rumen contents creating a stable foam which traps the gas;
- feedlot (grain) bloat—high concentrate, finely ground diets increase the acidity of the rumen fluid leading to decreased rumen motility; acid-tolerant bacteria produce excessive quantities of gas which cannot be expelled;
- free-gas bloat—the result of obstruction or failure to eructate (belch); can be caused by particles lodged in the esophagus, toxins interfering with the nervous system or as a result of physical defects that occur during organ or tissue development.

Possible remedies for bloat include:

- vigorous exercise to stimulate gas release;
- addition of froth dispersing agents (e.g., poloxalene) to feed (pasture bloat only);
- introduction of a stomach tube to remove free gas;
- in severe cases, insertion of a trocar and cannula through the animal's left side to allow gas to escape.

Displaced abomasum

A displaced abomasum blocks the flow of digesta to the small intestine. The form of displaced abomasum seen in 90% of cases is left displacement (LDA). It is common in early lactation—most LDAs are diagnosed in the first 2 weeks post-partum. Excess gas collects in the abomasum causing it to rotate under the rumen and then up the left side of the body behind the rumen, which is emptier than usual because of decreased feed intake. Surgical repositioning of the abomasum is often necessary but costly. Prevention is possible if the following risk factors can be controlled:

Decreased dry matter intake (DMI): in the weeks just prior to and after calving, reduced DMI can result in decreased rumen fill, allowing more room for the abomasum to displace (see The Goldilocks Strategy; sidebar, page 75) .

Ketosis: LDAs and ketosis are very closely related because intake is decreased in animals suffering from this metabolic disorder.

Calving-related disorders: retained placenta, uterine infections (metritis) and difficult calvings (dystocia) are often accompanied by decreased feed intake and rumen fill.

Milk fever: milk fever occurs when the cow does not have sufficient circulating calcium to produce milk and support regular body functions. Muscles require calcium to contract and maintain their tone. Hypocalcemia can reduce both rumen and abomasal tone and motility.

Ration physical form: diets containing too few coarse particles require less chewing which reduces saliva production. There is also a reduction in the fibre mat (page 44) and in rumen motility, both of which are strong risk factors in the incidence of LDAs.

Ruminal acidosis and laminitis: diets containing too little physically effective fibre can result in ruminal acidosis. Under these conditions, rumen bacteria produce toxins which enter the bloodstream. As blood moves through the laminae of the hoof, they become damaged and inflamed. The resulting lameness means the animal spends more time lying down and less time standing to eat, predisposing her to LDA.

Over or underconditioning: one study found that cows with body condition scores of 4 or higher (on a 1 - 5 scale) had a 15.7% incidence of LDA. Overconditioned cows also have an increased risk of developing ketosis and fatty liver, which further increase the risk of LDAs.

Transition cow environment: the novelty of new surroundings, the presence of dominant cows, the metabolic demands of increasing milk production

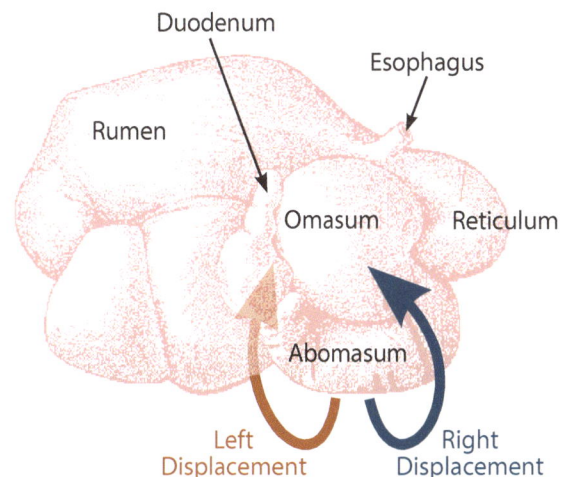

Figure 6.16: The abomasum (fourth compartment of the ruminant stomach) can rotate under the rumen and omasum to the left or right. Left displacements are, by far, the more common.

and a new ration can make the transition period a stressful time for both cows and heifers. Management programs which focus on cow comfort, proper ventilation and sufficient feedbunk and waterer space will reduce stress on fresh cows. Any practice which removes impediments to feed intake will decrease the risk of LDA.

Fat cow syndrome

Excessively fat cows are particularly susceptible to post-calving metabolic disorders, including retained placenta, displaced abomasum, ketosis and milk fever. In fat cow syndrome, these disorders are often associated with fatty liver. However, fatty liver may also develop in cows who are not excessively fat.

Cows are judged excessively fat when their condition score is greater than 4. These cows typically have reduced appetites both before and after calving. Low feed intake results in negative energy balance which may trigger the metabolic events leading to clinical disorders.

Fatty liver syndrome

In late pregnancy and early lactation, energy demands for fetal growth and milk production may exceed energy intake. In response, free fatty acids (FFA) are mobilized from body fat depots. Although some of these FFA may be used directly by tissues as a source of energy, most are absorbed by the liver. The liver itself can either derive energy from the FFA, or repackage them as triglycerides which it may store or transport to other tissues in the form of lipoproteins. In the lactating cow, the mammary gland uses triglycerides for milk fat synthesis.

Fatty liver develops when the rate of FFA absorption by the liver exceeds the rate at which triglycerides can be exported or oxidized. Overconditioned cows (condition score >4) are particularly susceptible. Fatty liver may develop pre-partum due to the increasing energy requirements of pregnancy combined with reduced feed intake in the last 7 - 10 days before calving or post-partum due to lactation energy demands.

Figure 6.17: The top piece of liver tissue has been infiltrated with fat. The lower piece is normal.

Grass tetany

Correctly named *hypomagnesemic tetany*, grass tetany results from a decrease in blood magnesium (Mg) concentration. The disorder is called grass tetany because it is most commonly observed in lactating cows grazing on rapidly-growing grass pastures having low Mg levels. Milk production places a high demand on blood Mg and, when feeds are low in Mg, blood levels fall.

High soil potassium (K) concentrations inhibit the uptake of Mg by plants. In some areas of western Canada (e.g., BC's Fraser Valley), large amounts of dairy manure applied to relatively small areas of land have resulted in high soil K levels. Forages harvested from these lands often have very low Mg concentrations and K levels in the 3 - 5% range.

In acute hypomagnesemia, affected cows suddenly throw up their heads, bellow, run aimlessly, fall, and exhibit paddling convulsions. In less severe cases, they are obviously ill at ease, walk stiffly, are hypersensitive to touch and sound and urinate frequently. It may take 2 - 3 days for signs to progress to the acute convulsive stage.

Today, grass tetany is seldom seen in western Canadian herds because so few lactating cows are permitted to graze. However, it may be more common among cows in confinement as high production levels place increasing demands on blood Mg.

Low blood Mg can inhibit the activity of parathyroid hormone which is involved in the release of calcium (Ca) from bone. Cows with hypomagnesemia often do not exhibit signs of a problem until blood Ca concentrations are low, as they often are at calving. Not infrequently, a fresh 'downer' cow who is assumed to be suffering from milk fever fails to respond to intravenous (IV) Ca. When this occurs, IV Mg + Ca (Cal-Mag) is recommended, as the primary cause may be hypomagnesemia.

Ketosis

Ketosis (hyperketonemia) is a metabolic disorder that usually occurs 10 days to 6 weeks after calving in high-producing cows. The initiating event is a drop in blood glucose concentration (hypoglycemia). Since the cow absorbs little glucose from her diet, most of that required for milk lactose synthesis must be manufactured by the liver, from propionic and amino acids (gluconeogenesis). High demand for lactose in early lactation and/or reduced availability of glucose precursors due to inadequate feed intake trigger the hypoglycemia. Low feed intake may be due to acidosis, milk fever, metritis or displaced abomasum.

In response to hypoglycemia, free fatty acids are mobilized from fat depots and taken up by the liver.

End-products of liver fat metabolism are the ketones: acetone, acetoacetic acid and β-hydroxybutyrate. These can be detected in blood, milk, urine and exhaled air. Since urine concentrations are usually 4 times those in blood and 8 times those in milk, a simple colorimetric urine test will detect elevated levels.

Animals afflicted with clinical ketosis go off-feed, become constipated and depressed, lose body condition and reduce milk production. A majority of high-producing cows likely experience some degree of subclinical ketosis in early lactation without exhibiting any obvious symptoms. Reduced feed intake in the last few days before calving can provoke fat mobilization and accumulation in the liver (see page 78) which may reduce its capacity for gluconeogenesis and predispose cows to ketosis immediately post-partum. Therefore, every effort should be made to correct the hypoglycemia in the early stages, before gluconeogenesis becomes severely impaired. Intravenous glucose may have some immediate benefit, but propylene glycol administered as a drench is the preferred treatment. Monensin supplementation in the close-up dry cow diet may also reduce the risk of ketosis.

Milk fever

The day she calves, a cow will commonly produce 10 litres of colostrum, containing about 23 grams of calcium (Ca). She will need another 23 grams for maintenance, resulting in a total requirement which is about 12 times the amount of Ca circulating in her blood at any point in time. Most cows respond by increasing the absorption of Ca from the ration, mobilizing Ca from bone and reducing urinary Ca excretion. Cows that don't respond effectively will end up with low blood calcium levels (hypocalcemia).

Because Ca is essential for muscle tone and contraction, low blood levels can result in cows going down with milk fever. Poor muscle tone also contributes to displaced abomasum (page 77). And weak uterine muscle contractions are involved in retained placenta (see below).

Strategies to prevent these problems are aimed at feeding the dry cow to prepare her systems for the increased Ca demand at calving. Limiting Ca intake in the close-up dry period has been commonly recommended, recognizing that high dietary calcium levels result in decreased absorption from the digestive tract. And, when higher Ca intakes keep blood Ca levels high, excretion increases and the hormonal mechanism for mobilizing Ca from bone is effectively shut down.

A second strategy for reducing the risk of milk fever uses the concept of dietary cation-anion difference (DCAD) which is described on pages 61 and 75.

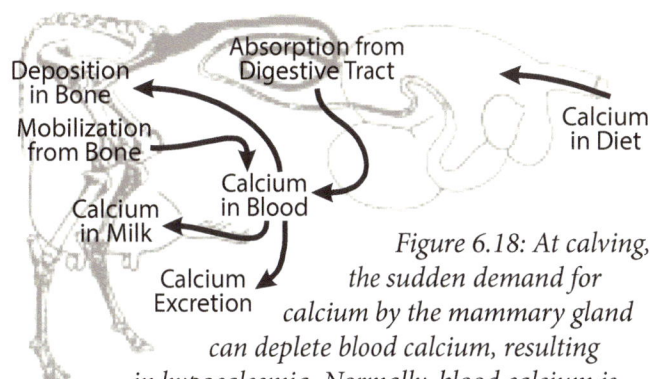

Figure 6.18: At calving, the sudden demand for calcium by the mammary gland can deplete blood calcium, resulting in hypocalcemia. Normally, blood calcium is replenished by calcium absorbed from the digestive tract and mobilized from bone. When this does not occur, blood calcium drops and milk fever develops.

Retained placenta

The cow's placenta is made up of numerous cotyledons which are attached to caruncles on the wall of the uterus (figure 6.19). These button-like structures are the sites of nutrient transfer from dam to fetus. Normally, the cotyledons detach from the caruncles within an hour or two after calving and the placenta is expelled from the uterus. If it has not been expelled after 24 hours, it is considered retained.

Retained placenta is usually attributed to one of the following causes:
• Early calving, 5 or more days before the due date;
• Dystocia (difficult calving);
• A reproductive tract infection causing fever, abortion or stillbirth;
• Dietary deficiency of vitamin A (or its precursor, beta carotene), vitamin E, selenium or iodine;
• High body condition score at calving (>4) which may be related to poor uterine muscle tone;

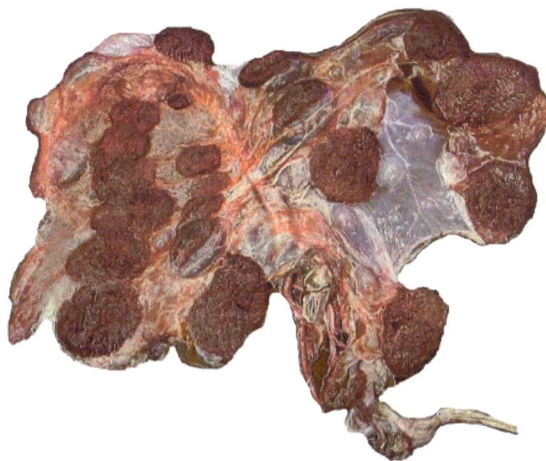

Figure 6.19: Before calving, the cotyledons (the dark red patches) on this placenta were attached to caruncles on the wall of the uterus, providing a pathway for nutrient transfer from dam to fetus.

• Uterine contractions which are too weak to expel the cotyledons from their attachment sites, the uterine caruncles. Weak contractions may be related to low blood calcium (hypocalcemia or subclinical milk fever).

Most cows with retained placenta will develop metritis, an infection of the uterus evident in a fetid red-brown watery or purulent vaginal discharge. Older cows have a higher risk of retained placenta than first parity cows.

Treatment of retained placenta is controversial. Most veterinarians recommend waiting 1 to 3 days before attempting to clean a cow manually, if at all. Some favour intravenous, intramuscular or subcutaneous antibiotics; few recommend intrauterine boluses.

Udder edema

Udder edema is indicated by swelling of the lower part of the udder due to excessive fluid accumulation between the milk secretory cells. Often the swelling extends forward, under the skin in front of the udder. In some cases edema is seen in the vulva and brisket areas. It is most common in the few weeks around calving (periparturient period) in primiparous cows (heifers) and tends to be more severe in heifers calving after 26 months of age and those with male calves.

Figure 6.20: Subcutaneous edema, showing swelling in front of the udder.

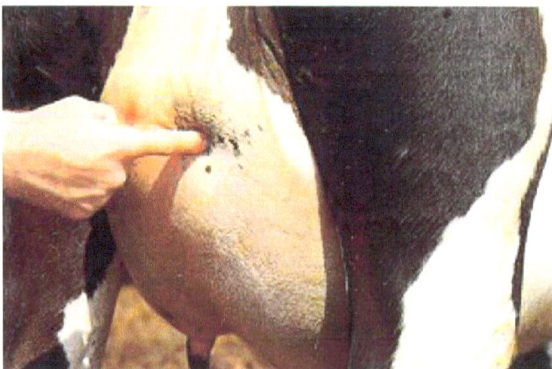

Figure 6.21: Finger pressure applied to an edematous udder creates a depression which persists for 30 - 60 seconds after the finger has been removed.

Overconditioned (condition score >4) older cows and those that previously experienced udder edema are also more susceptible.

The cause of udder edema is unknown, although there is considerable evidence to suggest that excessive sodium or potassium intake prepartum might be involved. Other causative factors that have been proposed include genetics or an imbalance of reproductive hormones pre-partum.

Effective treatment of cows with udder edema is limited to massaging swollen tissue in an upward direction for 15 - 20 minutes after each milking, in an attempt to help drain excess fluid into the general circulation. Diuretics may help by increasing fluid excretion. Some diuretic treatments also include corticosteroids which may reduce inflammation. Restricting water and salt intake may also be of some benefit.

Chapter 7: Ration Formulation

Ration formulation is the process used to determine an appropriate mix of available feeds that will satisfy animal nutrient requirements. This can only be done if feed nutrient concentrations, animal nutrient requirements and potential ration intake are known.

Dry matter intake

The first step in formulating a diet for any individual or group of animals is the prediction of maximum individual dry matter intake (DMI) of a ration offered *ad libitum*; for a group, an estimate of the group average DMI is required. It should be recognized at the outset that the amount of feed a cow will voluntarily consume is dependent on many factors, including:

- each individual animal's drive to eat which is affected by her genetics, age, size, production level, health, reproductive and social status, as well as other factors influenced by management, housing, facilities and environment;
- feed accessibility: the frequency and amount of feed offered and the animals' ability to access it;
- palatability—feed texture, moisture content and flavour;
- diet digestibility—dietary fibre (especially if it is heavily lignified; see pages 3 and 48) is more slowly digested that other dietary components resulting in slower passage rate through the rumen, reducing intake capacity;
- negative metabolic and health effects of the diet itself due to feed contaminants, toxins and excessive concentrations of essential nutrients (e.g., starch, degradable protein, minerals);
- water quality and accessibility—insufficient and poor-quality water have negative effects on voluntary feed intake.

Dry matter intake predictions for calves

Maximum DMI of calves on liquid diets is about 2% of body weight in the first week of life, rising to around 2.25% thereafter; up to 2.5% for calves with body weights greater than 65 kg. By the time calves reach 100 kg, their actual DMI should be in the range of 3 - 4% of body weight.

A prediction equation for starter DMI derived from 28 studies in the US and the Netherlands is as follows:

$$\text{Starter DMI (g/d)} = -652.525 + (\text{BW} \times 14.734) + (\text{MEiLD} \times 18.896) + (\text{FPstarter} \times 73.303) + (\text{FPstarter}^2 \times 13.496) - (29.614 \times \text{FPstarter} \times \text{MEiLD})$$

where:
- BW is body weight in kg;
- MEiLD is metabolizable energy intake from the liquid diet in Mcal/d;
- FPstarter is weeks since the first offer of starter.

Dry matter intake prediction for growing heifers

If ration NDF concentration is not known, the following equation is recommended to predict DMI for growing heifers:

$$\text{DMI, kg/d} = 0.022 \times \text{MatBW} \times (1 - e^{\{-1.54 \times (\text{BW/MatBW})\}})$$

If ration NDF concentration is known:

$$\text{DMI, kg/d} = [0.0226 \times \text{MatBW} \times (1 - e^{\{-1.47 \times (\text{BW/MatBW})\}})] - [0.082 \times (\text{NDF} - \{23.1 + 56 \times (\text{BW/MatBW}) - 30.6 \times (\text{BW/MatBW})^2\})]$$

where:
- BW is body weight in kg;
- MatBW is mature body weight in kg (use 700 kg for Holsteins; 520 for Jerseys);
- NDF is ration NDF concentration in % of DM.

Dry matter intake predictions for lactating and dry cows

Several DMI prediction formulas for lactating and dry dairy cattle have been proposed over the past many years. Based on the most recent research, two formulas are proposed in NASEM Dairy 8. The first is based on animal factors only and is the one used in the NASEM Dairy 8 computer model described in Appendix C:

$$\text{DMI (kg/d)} = [3.7 + \text{Parity} \times 5.7) + 0.305 \times \text{MilkE} + 0.022 \times \text{BW} + (-0.689 - 1.87 \times \text{Parity}) \times \text{BCS}] \times [1 - (0.212 + \text{Parity} \times 0.136) \times e^{(-0.053 \times \text{DIM})}]$$

where:
- Parity is 0 if all cows in the group are primiparous; 1 if they are all multiparous. If a mixed parity group, then Parity = multiparous cows / total cows.
- MilkE is milk net energy output in Mcal/day, calculated as follows:
  $$\text{MilkE (Mcal/d)} = [(0.0929 \times \text{MilkFat\%}) + (0.0585 \times \text{MilkTrueProtein\%}) + (0.0395 \times \text{MilkLactose\%})] \times \text{MilkYield (kg/d)};$$
- BCS is body condition score (on 1 – 5 scale);
- DIM is days in milk.

The second DMI prediction equation proposed in NASEM Dairy 8 is based on ration fibre and milk yield:

DMI (kg/d)= 12.0 - 0.107 x fNDF + 8.17 x ADF/NDF
+ 0.0253 x fNDFD – 0.328 x (ADF/NDF – 0.602)
x (fNDFD – 48.3) + 0.225 x MY
+ 0.00390 x (fNDFD – 48.3) x (MY – 33.1)

where:
· fNDF is % of dietary NDF contributed by forage;
· ADF/NDF is ADF as a fraction of NDF in the diet;
· fNDFD is % digestibility of forage NDF measured *in vitro* or *in situ*;
· MY is milk yield (kg/d).

These prediction equations can be used to calculate an initial estimate of DMI for the purpose of ration formulation. Once the formulated ration is fed, it is important to estimate actual intake and make adjustments to the ration and to feeding management in an attempt to maximize consumption. For lactating cows in particular, higher DMI drives milk yield but also makes it possible to increase the fibre:starch ratio, reducing the risk of ruminal acidosis.

The formulation process

The process of ration formulation involves finding a mathematical solution that satisfies a number of constraints—primarily, the proportions of each of the available ingredients that, when mixed together, will satisfy a set of feed fraction concentration targets in the complete product. That product may represent a complete daily diet, such as a 'total mixed ration' (TMR) or only a portion of the daily diet such as a protein or mineral-vitamin supplement.

Additional constraints may include:
• achieving a least-cost solution;
• restricting the inclusion rates of ingredients between minimum and maximum values;
• restricting the moisture level in the finished product;
• restricting mineral concentrations in the final product between a minimum requirement and a potentially toxic maximum level.

Ration formulation is an inexact science subject to:
1. uncertainty in the definition of nutrient requirements:
   • requirement recommendations are derived from feeding trials conducted in a variety of geographic locations under widely varying environmental conditions.
2. uncertainty in feed analysis values used, arising from:
   • samples submitted for analysis not being representative of the entire batch of feed being fed;
   • variation in the feeding value of various analytical fractions due to feed type, plant maturity and environmental growing conditions as well as error in the equations used to calculate nutrient concentrations from analytical fractions (e.g., digestible energy from ADF—see page 37);
   • permitted tolerances in the (as-fed) guaranteed analysis values of commercial feeds; for example, the Canadian *Feeds Regulations* (Schedule I, Table 1) permits a deficiency of 1 percentage point in crude protein concentration in manufactured feeds with a guaranteed analysis of 24% and under. So a 16% CP complete feed is permitted to contain only 15% CP.
3. variation in animal responses to formulated rations due to:
   • errors in mixing ration ingredients;
   • uncertainty in prediction of feed intake by individual animals;
   • variation in feed conversion efficiency of individual animals due to genetics, gender, stage of maturity and environment.

Recognizing these sources of uncertainty, ration formulation should only be considered the starting point for feeding a group of animals. After offering the formulated diet for a week to 10 days, animal reponses to the formulated ration should be assessed:
• is average consumption by the group in the range expected?
• is there evidence of unequal consumption among individuals?
• is there evidence of separation of mixed ration ingredients or sorting by the animals?

Balanced rations

Previous chapters have alluded to the importance of 'well-balanced' diets. The practical consequence of dietary balance is explained by the 'most limiting nutrient' concept which states that production is restricted to the level that can be supported by the most limiting nutrient. This is illustrated diagrammatically

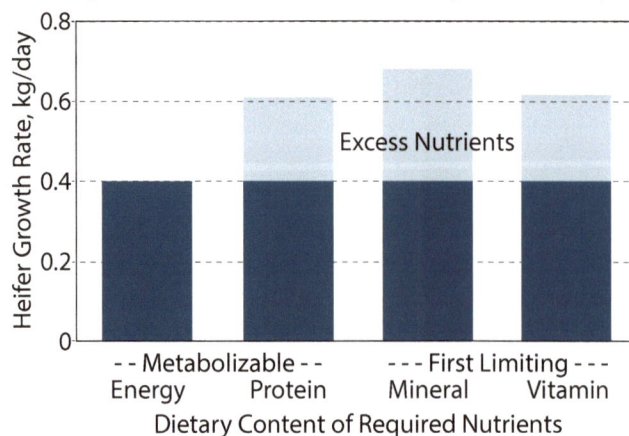

Figure 7.1: Production is limited to the level that can be supported by the most limiting nutrient. In this example, limited dietary metabolizable energy content limits heifer growth to 0.4 kg/day; other excess nutrients are wasted.

in figure 7.1. The net result of imbalanced rations is the inefficient use of feed resources.

Ration formulation software

Explanation of the mathematics involved in ration formulation is beyond the scope of this publication but several good computer programs are available which make formulation a simple matter. The Dairy Ration Formulator shown below illustrates a typical formulator layout. This one was built as a simple Excel® workbook where a solution can be determined manually or using Microsoft Excel's Solver® add-in. A copy of this workbook is available from the author on request. NASEM Dairy 8 includes the computerized formulation model which is described in Appendix C. More sophisticated formulation software is available that includes computation of nutrient requirements, feed libraries, ingredient inventory control, least-cost formulation capability, and many other features.

Although software can simplify the task of ration formulation, without a solid understanding of nutrition principles and some experience feeding dairy animals, completely unrealistic rations can also be computed.

Dairy Ration Formulator

MMC-200-43/8				Mature multiparous cow, 700 kg, 200 DIM, 43 kg Milk, 3.8% Fat, 3.3 % Protein									
	Body Weight	700	kg	ADG	0.5	kg/day	Milk	43	kg/day	BF	3.8	%	

Diet Ingredients	DM Inclusion Rate, %		DMI	As-fed	DM	DE	CP	RUP	NDF	Starch	Ca	P	Cost
	Min	Max	Solution	kg	kg	%	Mcal/kg	%	%	%	%	%	%	$/tonne wet
Barley Silage, mid maturity	0	100	27.8	7.50	20.76	36.1	2.62	11.4	3.1	52.9	12.3	0.46	0.31	45.00
Grass lg mixt, leg., hay immtr	0	100	7.4	2.00	2.26	88.6	2.64	20.4	4.4	43.9	1.9	1.28	0.30	125.00
Beet pulp, dry	0	100	3.7	1.00	1.08	92.3	2.74	9.9	7.0	46.9	0.6	0.77	0.08	245.00
Barley Grain, steam rolled	0	100	29.6	8.00	9.01	88.7	3.43	11.8	2.6	18.6	56.7	0.12	0.38	215.00
Corn grain, steam-flaked	0	100	7.4	2.00	2.33	85.7	3.62	8.0	5.5	8.6	71.7	0.06	0.24	265.00
Canola meal	0	100	9.3	2.50	2.80	89.1	3.10	41.5	13.3	29.0	1.6	0.79	1.15	210.00
DDGS, high protein	0	100	7.4	2.00	2.20	91.1	3.30	39.0	17.7	37.6	6.2	0.08	0.64	305.00
Blood meal, high dRUP	0	100	5.6	1.50	1.65	90.9	4.52	95.1	70.8	0.0	0.0	0.14	0.28	655.00
Calcium soaps	0	2	1.9	0.50	0.52	95.3	5.42	0.0	0.0	0.0	0.0	0.00	0.00	1,342.00
None	0	100	0.0	0.00	0.00	0.00	0.00	0.0	0.0	0.0	0.0	0.00	0.00	0.00
Requirement		100.0	27.4				2.60	17.5	7.5	25.0		0.58	0.35	
Maximum							3.50	18.0		33.0	30.0	0.60	0.40	
Solution		100.0	27.0	42.63	63.3		3.193	21.1	9.1	31.3	26.3	0.38	0.41	166.14
Excess/Deficit		0.0	(0.4)				0.59	3.6	1.6	6.3	(3.7)	(0.2)	0.1	

Figure 7.2: A typical ration formulator layout that defines the concentrations of feed fractions in ration ingredients, ingredient costs, animal feed fraction requirements and constraints on inclusion rates of each ingredient.

Appendix A: Feed Analysis Tables

Feed Name	DM % of wet	Ash % of DM	CP % of DM	A fraction % of CP	B fraction % of CP	C fraction % of CP	Kd of B fraction	RUP % of CP	RUP % digestible	Soluble CP % of CP
Alfalfa meal	90.7	11.7	19.5	28	66	6	6.7	36	75	37.0
Barley hay	89.4	8.6	10.8	57	33	10	7.0	27	83	44.8
Barley silage, headed	37.0	5.9	10.9	57	33	10	7.0	27	83	61.4
Barley silage, mid-maturity	36.1	7.9	11.4	57	33	10	7.0	27	83	59.9
Barley silage, vegetative	33.0	10.7	14.2	57	33	10	7.0	27	83	65.0
Cool season grass hay, mature	89.8	6.7	9.2	30	56	14	5.5	42	60	29.3
Cool season grass hay, mid-mature	88.3	8.6	13.3	30	56	14	5.5	42	60	30.5
Cool season grass silage	38.8	8.1	13.4	52	35	13	5.7	32	60	48.0
Corn silage, immature	31.3	4.0	7.9	58	26	16	4.0	34	70	52.7
Corn silage, mature	39.6	3.7	7.5	48	28	24	3.0	44	70	51.3
Corn silage, typical	35.4	3.8	7.7	60	24	16	4.1	33	70	51.8
Grass-leg mix, mostly grass, silage	39.6	8.1	14.3	41	49	10	13.0	26	70	49.2
Grass-leg mix, mostly grass, hay, mid	89.2	9.4	15.6	41	49	10	13.0	26	70	34.2
Grass-leg mix, mostly grass, hay, mature	87.4	7.5	10.9	41	49	10	13.0	26	70	29.7
Grass-leg mix, mostly leg, hay, mature	85.3	9.0	17.4	34	51	15	9.5	34	65	32.0
Grass-leg mix, mostly leg, hay, immature	88.6	10.0	20.4	44	49	7	15.1	22	75	37.2
Grass-leg mix, mostly leg, silage	41.0	10.3	20.0	52	39	9	8.2	27	72	53.6
Grass-leg mix, mix, hay	86.4	9.2	12.1	41	49	10	13.0	26	70	27.0
Grass-leg mix, mix, silage	40.0	9.5	17.7	61	30	9	10.6	22	70	50.7
Legume hay, immature	89.3	11.0	21.5	43	50	7	17.8	20	65	41.1
Legume hay, mature	87.7	10.0	18.1	39	49	12	14.0	27	65	37.9
Legume hay, mid-maturity	88.1	10.8	20.7	45	46	9	17.8	22	65	34.6
Legume silage, immature	41.6	11.1	22.1	62	29	9	13.1	21	70	55.2
Legume silage, mid-maturity	42.9	10.6	20.5	52	39	9	8.2	27	70	49.3
Millet hay	87.5	11.2	10.7	28	53	19	5.0	47	60	34.7
Millet silage	29.3	11.7	13.0	38	29	33	3.7	52	55	47.0
Oat hay mid-maturity	89.5	7.4	8.5	35	53	12	4.3	42	70	39.5
Oat silage, immature	33.7	13.4	18.5	46	30	24	5.4	41	65	57.6
Oat silage, mid-maturity	35.8	10.4	12.9	45	31	24	5.4	42	65	58.9
Pasture, grass	19.3	9.6	25.1	57	33	10	6.0	28	65	25.4
Pasture, grass-legume mixture	19.3	9.6	25.1	57	33	10	6.0	28	65	25.4
Pasture, legume	13.9	9.6	26.1	62	29	9	13.1	21	70	55.2
Pea hay	89.5	9.1	15.9	45	46	9	17.8	22	65	45.9
Pea silage	31.7	11.4	17.0	52	39	9	8.2	27	72	59.0
Rye annual fresh, immature	15.8	10.9	27.5	57	33	10	6.0	28	65	25.4
Rye annual fresh, mid-maturity	19.5	9.8	20.5	57	33	10	6.0	28	65	28.6
Rye annual hay, immature	90.0	11.1	22.9	57	33	10	6.0	28	65	39.2
Rye annual hay, mature	92.7	6.3	7.6	57	33	10	6.0	28	65	36.0
Rye annual hay, mid-maturity	90.3	9.3	12.0	57	33	10	5.9	29	65	41.0
Rye annual silage, immature	34.7	11.1	16.4	57	33	10	5.9	29	65	65.7
Rye annual silage, mature	67.2	8.7	8.3	57	33	10	5.9	29	65	46.8
Rye annual silage, mid-maturity	34.9	10.3	14.4	57	33	10	5.9	29	65	61.7
Triticale hay	91.0	8.5	10.3	56	33	11	5.9	30	65	46.8
Triticale plus pea silage	34.2	10.3	16.0	56	33	11	5.9	30	65	62.6
Triticale silage, mature	30.1	10.1	14.2	56	33	11	5.9	30	65	62.7
Triticale silage, mid-maturity	33.3	12.4	17.8	56	33	11	5.9	30	65	66.8
Wheat hay, headed	90.6	8.1	9.9	35	53	12	4.3	42	70	39.8
Wheat hay, vegetative	90.5	8.2	10.5	35	53	12	4.3	42	70	41.6
Wheat silage, headed	34.8	10.5	10.7	62	29	9	10.0	22	72	69.5
Wheat silage, vegetative	35.6	9.9	13.4	62	29	9	10.0	22	72	65.0
Wheat straw	89.5	8.0	4.5	10	51	39	1.4	80	62	36.0

Table A1.1: Typical analyses of forages commonly fed to western Canadian dairy cattle.

Feed Name	ADIP % of DM	NDIP % of DM	ADF % of DM	NDF % of DM	IVdNDF48 % of NDF	Lignin % of DM	Starch % of DM	WS Carb % of DM	Fatty Acids % of DM	Crude Fat % of DM	DE Base Mcal/kg
Alfalfa meal	1.62	3.97	33.9	42.9	49.1	7.55	1.8	8.3	1.61	2.35	2.50
Barley hay	0.66	2.24	34.1	54.5	61.3	4.07	7.3	14.8	1.41	2.74	2.62
Barley silage, headed	0.95	1.30	26.3	44.8	50.0	3.81	23.9	7.0	2.06	3.08	2.86
Barley silage, mid-maturity	1.11	1.60	34.1	52.9	52.1	4.96	12.3	5.9	1.70	3.13	2.62
Barley silage, vegetative	1.39	2.00	37.7	56.6	57.5	5.19	3.0	6.9	2.07	3.65	2.52
Cool season grass hay, mature	1.48	3.79	41.4	66.7	55.8	5.97	2.0	10.8	0.96	2.35	2.34
Cool season grass hay, mid-mature	1.43	4.85	35.5	58.0	67.8	4.17	2.2	15.2	1.58	3.23	2.53
Cool season grass silage	1.60	3.69	39.0	62.1	63.6	5.82	1.9	7.3	1.84	3.63	2.44
Corn silage, immature	0.84	1.26	25.5	42.6	53.4	3.15	30.2	2.9	2.32	2.96	2.93
Corn silage, mature	0.80	1.19	23.2	39.3	50.8	2.97	35.5	3.1	2.36	2.86	2.88
Corn silage, typical	0.82	1.23	24.3	40.9	52.0	3.05	32.9	3.0	2.35	2.92	2.93
Grass-leg mix, mostly grass, silage	1.61	3.51	37.0	57.7	55.2	5.62	2.6	8.0	1.98	3.75	2.55
Grass-leg mix, mostly grass, hay, mid	0.99	1.33	33.8	54.7	67.5	4.28	1.5	12.6	1.93	3.39	2.64
Grass-leg mix, mostly grass, hay, mature	1.31	3.57	40.0	62.0	47.5	6.08	2.5	10.9	1.29	2.43	2.43
Grass-leg mix, mostly leg, hay, mature	1.83	4.38	38.7	51.2	20.1	8.27	2.5	5.0	1.23	2.27	2.40
Grass-leg mix, mostly leg, hay, immature	1.62	4.25	32.9	43.9	49.9	6.92	1.9	9.0	1.78	2.63	2.64
Grass-leg mix, mostly leg, silage	1.99	3.54	33.9	45.9	56.2	6.60	2.1	6.6	1.99	3.81	2.60
Grass-leg mix, mix, hay	1.66	4.12	39.3	58.2	54.5	6.70	2.7	10.1	1.37	2.66	2.38
Grass-leg mix, mix, silage	1.91	3.70	34.8	51.2	61.1	5.90	2.1	8.1	2.03	4.04	2.60
Legume hay, immature	1.51	2.80	30.7	37.7	51.4	6.59	2.3	9.8	1.54	2.55	2.68
Legume hay, mature	1.75	3.89	37.2	46.6	43.4	8.12	2.3	9.8	1.21	2.25	2.45
Legume hay, mid-maturity	0.74	1.85	32.1	41.1	52.4	6.64	1.5	9.0	1.50	2.08	2.63
Legume silage, immature	1.60	2.68	32.0	38.7	53.3	6.55	1.9	7.3	1.98	3.22	2.70
Legume silage, mid-maturity	1.25	2.25	33.7	43.2	49.4	7.42	2.0	6.3	2.32	2.87	2.59
Millet hay	0.91	3.49	39.7	61.9	64.0	5.64	2.9	8.4	1.09	1.95	2.25
Millet silage	1.13	3.36	39.2	59.7	61.2	5.48	3.5	6.2	1.47	2.79	2.25
Oat hay mid-maturity	1.03	2.05	37.5	59.0	56.0	4.71	4.1	17.8	1.45	2.36	2.51
Oat silage, immature	1.09	2.94	31.7	45.8	59.3	3.89	1.8	4.3	2.24	4.23	2.57
Oat silage, mid-maturity	1.21	2.08	38.8	57.4	54.1	5.39	3.2	6.7	1.77	3.64	2.41
Pasture, grass	1.20	5.20	21.8	41.8	86.0	3.35	2.1	14.6	2.32	4.10	2.91
Pasture, grass-legume mixture	1.20	5.20	21.8	41.8	86.0	3.35	2.1	14.6	2.32	4.10	2.94
Pasture, legume	1.60	2.68	25.9	32.4	65.4	5.19	3.1	8.7	2.00	3.50	2.95
Pea hay	1.48	3.60	32.0	43.4	58.5	5.78	8.6	9.0	1.69	2.98	2.66
Pea silage	2.23	3.86	37.1	52.5	57.7	6.42	3.4	4.5	1.68	3.80	2.45
Rye annual fresh, immature	1.42	6.84	24.8	42.9	83.6	3.14	4.9	5.8	3.07	5.03	2.93
Rye annual fresh, mid-maturity	1.06	5.09	28.2	49.1	85.0	3.70	4.9	7.7	2.44	4.37	2.77
Rye annual hay, immature	1.69	5.69	27.2	47.0	85.5	3.44	2.0	12.8	2.53	4.46	2.79
Rye annual hay, mature	0.56	1.89	42.7	66.8	45.9	6.21	2.0	12.7	1.01	1.77	2.36
Rye annual hay, mid-maturity	1.09	2.50	36.7	57.3	59.8	4.84	2.2	14.0	1.47	2.89	2.50
Rye annual silage, immature	1.06	2.11	33.1	50.5	34.0	3.76	1.7	14.0	1.95	4.10	2.64
Rye annual silage, mature	0.70	1.17	42.9	66.2	66.3	5.71	1.4	12.3	1.25	2.20	2.32
Rye annual silage, mid-maturity	1.23	2.05	38.3	58.0	61.9	4.93	1.5	8.8	1.68	3.88	2.48
Triticale hay	0.51	1.88	38.3	60.0	63.7	4.84	3.1	14.0	1.46	2.29	2.48
Triticale plus pea silage	2.17	5.70	37.1	55.7	65.1	5.46	2.8	6.0	2.13	3.76	2.50
Triticale silage, mature	0.81	1.69	37.2	58.6	58.5	4.34	1.7	10.1	2.48	3.47	2.55
Triticale silage, mid-maturity	1.01	2.12	34.8	52.2	57.5	4.31	1.5	7.5	2.38	4.08	2.56
Wheat hay, headed	5.93	11.90	33.6	52.8	57.2	4.91	12.2	9.6	1.01	2.09	2.52
Wheat hay, vegetative	6.38	15.99	36.1	58.0	59.3	4.79	3.4	17.3	0.89	2.19	2.48
Wheat silage, headed	1.17	1.55	35.1	51.1	61.4	4.99	13.0	6.5	1.53	3.06	2.51
Wheat silage, vegetative	1.14	1.91	37.0	56.6	59.0	4.79	2.5	9.2	1.42	3.50	2.53
Wheat straw	1.26	1.51	53.1	76.9	41.8	8.19	1.8	4.2	0.55	1.49	1.96

Table A1.2: Typical analyses of forages commonly fed to dairy cattle in western Canada.

Feed Name	Ca	P	Mg	K	Na	Cl	S	Cu	Fe	Mn	Zn	Mo	Se
				% of DM							mg/kg		µg/kg
Alfalfa meal	1.50	0.27	0.30	2.33	0.12	0.67	0.26	8.83	951	50	23	1.71	400
Barley hay	0.38	0.25	0.17	1.96	0.38	0.95	0.17	6.78	422	36	30	1.32	100
Barley silage, headed	0.32	0.29	0.18	1.57	0.11	0.47	0.18	5.96	172	31	26	0.77	100
Barley silage, mid-maturity	0.46	0.31	0.19	2.00	0.18	0.65	0.18	7.88	272	40	30	1.43	100
Barley silage, vegetative	0.54	0.35	0.19	2.72	0.16	0.87	0.21	9.24	663	48	32	1.40	120
Cool season grass hay, mature	0.44	0.21	0.20	1.63	0.06	0.58	0.15	8.33	196	93	26	1.53	60
Cool season grass hay, mid-maturity	0.48	0.28	0.23	2.27	0.11	0.78	0.20	9.21	217	86	27	1.69	60
Cool season grass silage	0.55	0.31	0.21	2.29	0.08	0.67	0.20	9.46	450	95	33	1.87	60
Corn silage, immature	0.24	0.24	0.17	1.05	0.02	0.27	0.11	6.34	166	30	28	1.11	40
Corn silage, mature	0.23	0.23	0.16	0.92	0.03	0.24	0.10	5.91	146	28	26	1.11	40
Corn silage, typical	0.24	0.23	0.17	0.99	0.03	0.26	0.11	6.22	165	30	27	1.11	40
Grass-leg mix, mostly grass, silage	0.55	0.30	0.23	2.40	0.13	0.82	0.19	10.66	395	87	32	2.77	90
Grass-leg mix, mostly grass, hay, mid	0.63	0.33	0.27	2.34	0.08	0.65	0.22	0.52	10	295	83	29.65	90
Grass-leg mix, mostly grass, hay, mature	0.51	0.22	0.22	1.55	0.07	0.48	0.14	8.75	261	89	27	1.54	90
Grass-leg mix, mostly leg, hay, mature	1.24	0.27	0.28	2.52	0.06	0.44	0.21	10.29	421	56	26	1.81	90
Grass-leg mix, mostly leg, hay, immature	1.28	0.30	0.30	2.25	0.07	0.56	0.24	10.24	285	44	26	1.86	90
Grass-leg mix, mostly leg, silage	1.26	0.35	0.27	2.77	0.05	0.57	0.24	10.94	473	57	30	1.69	
Grass-leg mix, 50:50 mix, hay	0.74	0.25	0.21	1.91	0.04	0.43	0.03	9.98	309	65	28	1.91	
Grass-leg mix, 50:50 mix, silage	0.87	0.34	0.25	2.54	0.06	0.59	0.23	10.73	451	72	32	1.79	100
Legume hay, immature	1.51	0.29	0.32	2.49	0.20	0.76	0.19	10.23	430	43	26	2.75	200
Legume hay, mature	1.37	0.28	0.29	2.34	0.11	0.66	0.12	9.71	380	44	24	2.23	180
Legume hay, mid-maturity	1.40	0.28	0.32	2.39	0.21	0.77	0.18	9.69	421	38	26	2.54	190
Legume silage, immature	1.30	0.35	0.33	2.79	0.16	0.80	0.22	11.44	685	62	30	3.51	180
Legume silage, mid-maturity	1.25	0.35	0.30	2.82	0.12	0.64	0.14	10.59	534	55	29	2.09	180
Millet hay	0.54	0.29	0.31	2.71	0.03	1.07	0.18	9.12	286	105	43	1.58	
Millet silage	0.55	0.33	0.34	2.88	0.06	1.05	0.20	11.66	491	110	46	2.08	
Oat hay mid-maturity	0.32	0.21	0.14	1.71	0.42	0.94	0.13	6.77	257	66	20	1.37	
Oat silage, immature	0.72	0.40	0.25	3.18	0.30	1.19	0.26	10.03	1,031	115	34		
Oat silage, mid-maturity	0.51	0.34	0.20	2.68	0.23	0.90	0.19	8.42	611	70	30	1.47	
Pasture, grass	0.48	0.49	0.29	3.17	0.40	1.20	0.40	10.00	275	75	36	0.00	70
Pasture, grass-legume mixture	0.90	0.40	0.30	3.00	0.30	1.00	0.30	10.70	480	68	33	0.00	130
Pasture, legume	1.30	0.35	0.33	2.79	0.16	0.80	0.22	11.44	685	62	30	0.00	180
Pea hay	1.04	0.28	0.28	2.05	0.11	0.55	0.20	9.27	1,169	45	32	2.20	
Pea silage	0.88	0.34	0.24	2.86	0.05	0.72	0.21	11.06	1,205	56	35	1.57	
Rye annual fresh, immature	0.48	0.49	0.29	3.17	0.15	1.00	0.40	8.00	400	90	35	0.00	70
Rye annual fresh, mid-maturity	0.51	0.39	0.27	2.64	0.15	1.00	0.30	8.00	400	90	35	0.00	90
Rye annual hay, immature	0.59	0.40	0.26	3.17	0.51	1.46	0.29	9.43	406	113	37	1.49	90
Rye annual hay, mature	0.37	0.18	0.16	1.42	0.13	0.69	0.13	5.67	175	115	25	1.24	90
Rye annual hay, mid-maturity	0.51	0.28	0.20	2.31	0.26	1.09	0.17	8.14	365	90	30	1.42	90
Rye annual silage, immature	0.57	0.42	0.23	3.63	0.16	1.14	0.26	10.39	709	72	36	0.00	90
Rye annual silage, mature	0.36	0.25	0.16	2.07	0.08	0.85	0.15	6.68	396	89	27	1.60	90
Rye annual silage, mid-maturity	0.48	0.37	0.19	2.92	0.15	0.92	0.21	9.53	493	75	34	1.58	90
Triticale hay	0.33	0.24	0.15	1.99	0.04	0.77	0.15	6.26	184	38	25	1.44	
Triticale plus pea silage	0.68	0.35	0.21	2.76	0.08	0.71	0.20	8.79	666	45	28	1.22	
Triticale silage, mature	0.38	0.34	0.17	2.82	0.05	0.86	0.20	8.91	455	48	35	1.46	
Triticale silage, mid-maturity	0.52	0.41	0.19	3.42	0.06	1.11	0.24	11.44	616	57	42	1.66	
Wheat hay, headed	0.31	0.23	0.14	1.65	0.05	0.62	0.16	7.52	303	51	25	1.65	
Wheat hay, vegetative	0.29	0.21	0.14	1.75	0.07	0.69	0.16	7.87	403	59	27	1.50	
Wheat silage, headed	0.30	0.29	0.13	2.03	0.06	0.74	0.17	7.69	543	46	28	1.58	
Wheat silage, vegetative	0.43	0.32	0.17	2.53	0.07	0.83	0.19	9.01	625	63	32	1.68	
Wheat straw	0.38	0.12	0.12	1.41	0.08	0.49	0.11	6.20	224	50	17	1.44	

Table A1.3: Typical mineral analyses of forages commonly fed to dairy cattle in western Canada.

Feed Name	DM % of wet	Ash % of DM	CP % of DM	A fraction % of CP	B fraction % of CP	C fraction % of CP	Kd of B fraction	RUP % of CP	RUP % digestible	Soluble CP % of CP
Apple pomace or byproduct, wet	18.2	2.9	6.4	42	53	5	7.4	30	80	19.7
Bakery byproduct	90.1	4.4	12.8	48	44	8	16.2	22	90	18.0
Bakery byproduct, bread waste	67.4	3.6	14.9	48	44	8	16.2	22	90	20.6
Bakery byproduct, cereal	91.1	3.5	9.2	34	62	4	20.0	19	75	18.2
Bakery byproduct, cookies	89.3	4.6	12.9	48	44	8	16.2	22	90	22.3
Barley grain, dry, ground	88.7	2.8	11.8	30	61	9	22.7	22	85	25.3
Barley grain, steam-rolled	88.7	2.8	11.8	30	61	9	22.7	22	85	25.3
Barley malt sprouts	81.2	5.9	23.9	47	45	8	13.3	24	64	39.9
Beet pulp, dry	92.3	5.2	9.9	5	90	5	2.0	71	80	20.6
Beet pulp, dry, molasses added	90.6	7.5	8.9	5	90	5	2.0	71	80	19.3
Beet pulp, wet	22.7	7.3	9.1	5	90	5	2.0	71	80	16.3
Blood meal, high dRUP	90.9	3.3	95.1	10	61	29	1.9	75	85	19.5
Blood meal, low dRUP	90.9	3.3	95.1	10	61	29	1.9	75	65	19.5
Brewers grains, dry	93.1	4.6	25.3	18	67	15	4.5	52	74	17.7
Brewers grains, wet	22.5	4.6	28.1	47	44	9	3.9	37	83	11.2
Brewers yeast, dry	92.8	8.6	50.7	9	91	0	2.4	63	93	45.3
Brewers yeast, wet	13.4	6.4	43.3	9	91	0	2.4	63	93	59.4
Calcium fatty acid soaps	95.3	15.5	0.0							
Candy (not chocolate) byproduct	90.1	1.0	2.4	74	26	0	3.2	21	90	26.2
Candy byproduct, high protein	88.9	5.6	14.6	74	26	0	3.2	21	90	26.9
Canola meal	89.1	7.9	41.5	22	71	7	10.5	32	74	25.0
Canola seed, ground	95.0	4.5	23.4	35	60	5	20.1	20	50	49.3
Chocolate byproduct	94.3	2.6	10.0	74	26	0	3.2	21	90	27.1
Corn germ	90.2	5.9	15.4	41	45	14	10.0	32	73	35.4
Corn germ meal	90.2	3.7	26.1	14	50	36	12.0	52	73	24.5
Corn gluten feed, dry	89.2	7.5	23.2	51	39	10	7.0	30	79	45.7
Corn gluten feed, wet	45.6	7.2	23.1	51	39	10	7.0	30	79	57.0
Corn gluten meal	90.5	2.8	68.5	8	72	20	2.5	69	92	7.5
Corn grain dry, coarse grind	86.9	1.5	8.5	23	70	7	5.4	43	73	22.1
Corn grain dry, fine grind	86.9	1.5	8.5	23	70	7	5.4	43	73	22.1
Corn grain dry, medium grind	86.9	1.5	8.5	23	70	7	5.4	43	73	22.1
Corn grain HM, coarse grind	72.3	1.6	8.5	28	71	1	5.1	39	90	33.1
Corn grain HM, fine grind	72.3	1.6	8.5	28	71	1	5.1	39	90	33.1
Corn grain, steam-flaked	85.7	1.3	8.0	2	82	16	3.0	68	90	13.9
Cottonseed meal	89.9	7.6	46.7	25	56	19	7.2	44	83	14.5
Cottonseed, whole	91.4	4.2	23.3	45	48	7	14.8	23	74	26.6
Distillers grains with solubles	89.1	5.4	30.2	26	62	12	5.0	46	75	15.7
Distillers grains, high protein	91.1	4.0	39.0	26	62	12	5.0	46	75	16.6
Distillers grains, low fat	89.9	5.3	31.0	26	62	12	5.0	46	75	20.1
Distillers grains, modified	49.2	5.6	30.3	26	62	12	5.0	46	75	21.9
Distillers grains, wet	33.2	4.5	31.5	26	62	12	5.0	46	75	16.4
Distillers solubles	31.2	11.1	22.6	26	62	12	5.0	46	75	69.9
Feather meal	92.9	2.3	90.6	24	31	45	0.8	73	68	9.9
Fish meal	92.0	21.1	69.2	36	38	26	1.9	56	76	22.4
Flaxseed	92.7	4.0	22.8	18	67	15	5.4	49	84	43.1
Flaxseed meal	89.6	6.8	38.5	18	67	15	5.4	49	84	39.0
Fruit and vegetable byproduct, wet	19.8	8.4	13.6	42	53	5	7.4	30	80	41.0
Grain screenings, source unknwn	89.2	6.0	16.2	23	70	7	5.4	43	73	40.4
Meat and bone meal, porcine	96.1	26.2	56.6	32	42	26	8.8	44	61	14.9
Molasses	65.4	16.0	9.3	74	26	0	3.2	21	100	95.9

Table A2.1a: Typical analyses of concentrate ingredients commonly included in western Canadian dairy cattle diets.

Feed Name	DM % of wet	Ash % of DM	CP % of DM	A fraction % of CP	B fraction % of CP	C fraction % of CP	Kd of B fraction	RUP % of CP	RUP % digestible	Soluble CP % of CP
Oat grain	89.3	3.2	12.2	72	20	8	26.4	16	72	27.1
Oat hulls	91.4	5.8	5.0	10	51	39	1.4	80	62	36.7
Peas	88.9	3.6	24.3	57	42	1	16.0	15	89	75.1
Potato byproduct	23.0	5.6	10.0	5	90	5	2.0	71	80	39.1
Poultry byproduct meal	95.5	14.2	65.6	5	90	5	2.0	71	90	28.2
Rye grain	86.0	2.6	11.8	31	54	15	19.1	29	88	33.0
Safflower meal	94.0	4.9	26.2	23	71	6	10.4	31	75	31.2
Soybean hulls	90.4	5.2	11.9	26	70	4	6.4	37	68	26.3
Soybean meal, expeller	91.2	6.7	47.6	9	91	0	2.4	63	93	14.7
Soybean meal, extruded	93.4	5.7	40.4	18	80	2	8.7	33	91	17.4
Soybean meal, solvent 48CP	89.3	7.2	52.6	18	80	2	9.0	33	91	23.1
Soybeans, whole raw	89.1	5.3	40.0	26	74	0	9.3	28	90	42.1
Soybeans, whole roasted	94.0	5.6	40.0	18	77	5	9.3	34	87	15.6
Sunflower meal	90.2	7.4	37.0	42	53	5	29.2	16	90	27.3
Sunflower seed	92.7	3.5	20.1	66	32	2	17.0	14	80	50.5
Sweet corn cannery waste	22.5	5.1	9.8	30	68	2	5.0	39	61	50.7
Tallow	99.8	0.0	0.0							
Tomato pomace	24.7	5.5	19.3	42	53	5	7.4	30	80	38.0
Triticale grain	88.4	2.1	12.1	31	54	15	19.1	29	88	31.3
Vegetable oils	99.0	0.0	0.0							0.0
Wheat bran	90.1	5.5	17.4	43	51	6	24.2	18	69	40.0
Wheat grain	85.7	2.1	13.5	31	54	15	19.1	29	88	28.4
Wheat middlings	88.3	5.9	19.1	48	44	8	16.2	22	57	40.1
Whey, dry	92.6	9.7	17.8	90	10	0	5.0	11	95	91.1
Whey, wet	22.9	13.5	7.4	90	10	0	5.0	11	95	53.7

Table A2.1b: Typical analyses of concentrate ingredients commonly included in western Canadian dairy cattle diets.

Feed Name	ADIP % of DM	NDIP % of DM	ADF % of DM	NDF % of DM	IVdNDF48 % of NDF	Lignin % of DM	Starch % of DM	WS Carb % of DM	Fatty Acids % of DM	Crude Fat % of DM	DE Base Mcal/kg
Apple pomace or byproduct, wet	1.98	3.38	38.6	45.7	66.3	15.95	3.5	25.5	1.88	5.97	2.28
Bakery byproduct	0.46	1.06	5.6	12.7	52.0	1.69	51.7	19.0	7.68	8.68	3.59
Bakery byproduct, bread waste	0.62	1.04	3.2	6.3	52.0	1.29	53.5	11.4	4.76	5.76	3.66
Bakery byproduct, cereal	1.07	1.53	3.1	7.2	52.0	1.49	47.6	28.1	1.95	2.95	3.44
Bakery byproduct, cookies	1.13	2.11	7.6	14.4	52.0	2.17	36.2	18.2	9.04	10.04	3.58
Barley grain, dry, ground	0.70	1.38	7.3	18.6	51.5	1.72	56.7	4.9	1.31	2.31	3.36
Barley grain, steam-rolled	0.70	1.38	7.3	18.6	51.5	1.72	56.7	4.9	1.31	2.31	3.43
Barley malt sprouts	0.86	5.13	18.3	40.5	63.0	2.61	8.6	15.9	1.46	2.73	3.11
Beet pulp, dry	1.30	5.47	28.2	46.9	79.0	3.88	0.6	5.7	0.63	1.06	2.74
Beet pulp, dry, molasses added	1.79	4.23	26.1	39.7	74.0	3.53	1.1	13.8	0.63	1.19	2.73
Beet pulp, wet	0.88	2.22	27.2	44.1	78.3	3.18	0.9	3.0	0.64	0.99	2.73
Blood meal, high dRUP	4.25	5.67	0.0	0.0	0.0	0.00	0.0	0.0	1.31	1.69	4.52
Blood meal, low dRUP	4.25	5.67	0.0	0.0	0.0	0.00	0.0	0.0	1.31	1.69	3.72
Brewers grains, dry	3.12	8.60	24.7	51.8	51.5	6.66	6.5	3.8	8.31	9.03	2.95
Brewers grains, wet	2.94	3.96	23.8	49.3	47.3	6.64	5.2	2.0	7.61	9.52	3.10
Brewers yeast, dry	0.56	2.73	3.6	1.6	50.0	1.82	4.1	4.1	2.34	1.11	4.00
Brewers yeast, wet	0.48	2.33	5.6	11.5	50.0	1.25	4.4	14.0	2.34	3.34	3.84
Calcium fatty acid soaps	0.00	0.00	0.0	0.0		0.00	0.0		84.50	84.50	5.42
Candy (not chocolate) byproduct	0.46	0.84	1.5	2.3	50.0	0.57	23.5	60.0	0.25	1.25	3.48
Candy byproduct, high protein	2.76	3.97	19.5	29.7	50.0	5.16	16.1	15.0	11.11	12.11	3.31
Canola meal	2.50	4.75	20.3	29.0	49.4	8.51	1.6	11.0	2.51	3.51	3.10
Canola seed, ground	2.80	4.53	20.2	28.7	45.0	6.01	2.4	6.0	39.46	40.46	4.28
Chocolate byproduct	1.01	1.92	8.3	13.2	40.0	1.43	11.2	39.6	20.68	21.68	4.04
Corn germ	0.78	2.67	10.1	27.0	74.0	2.69	27.6	3.0	16.89	17.89	3.62
Corn germ meal	4.15	9.88	15.2	44.8	74.0	2.92	19.4	3.9	2.11	3.11	3.10
Corn gluten feed, dry	4.44	9.88	11.5	35.7	74.1	2.31	15.5	5.8	3.38	3.91	3.19
Corn gluten feed, wet	0.89	2.04	12.1	36.9	76.7	1.89	15.3	4.0	3.09	3.81	3.22
Corn gluten meal	1.11	1.96	3.7	6.8	73.0	1.79	16.4	1.6	1.44	2.44	4.31
Corn grain dry, coarse grind	0.52	0.91	3.6	9.8	62.3	1.37	70.4	2.9	3.84	3.85	3.10
Corn grain dry, fine grind	0.52	0.91	3.6	9.8	62.3	1.37	70.4	2.9	3.84	3.85	3.54
Corn grain dry, medium grind	0.52	0.91	3.6	9.8	62.3	1.37	70.4	2.9	3.84	3.85	3.45
Corn grain HM, coarse grind	0.44	0.77	3.5	9.6	50.9	1.21	70.9	3.0	3.57	3.58	3.52
Corn grain HM, fine grind	0.44	0.77	3.5	9.6	50.9	1.21	70.9	3.0	3.57	3.58	3.70
Corn grain, steam-flaked	0.63	1.29	3.4	8.6	55.7	1.26	71.7	1.8	3.14	3.14	3.62
Cottonseed meal	1.72	2.22	19.2	28.1	25.0	7.01	1.1	8.5	3.06	3.60	3.29
Cottonseed, whole	2.73	2.84	38.6	50.6	12.8	11.21	0.8	4.0	18.26	18.62	3.14
Distillers grains with solubles	2.85	3.83	14.6	32.1	71.5	4.17	4.5	4.6	11.39	12.55	3.46
Distillers grains, high protein	3.97	4.45	17.7	37.6	62.7	5.83	6.2	5.4	6.56	7.56	3.30
Distillers grains, low fat	3.15	3.87	14.8	30.8	47.2	3.53	6.1	8.0	7.90	8.90	3.41
Distillers grains, modified	4.09	4.69	14.4	27.1	52.8	3.09	4.7	9.7	8.35	9.35	3.47
Distillers grains, wet	3.29	4.13	16.1	31.7	25.5	3.22	6.4	7.3	8.31	9.31	3.47
Distillers solubles	0.78	1.21	3.2	4.8	70.0	0.57	4.0	28.7	9.99	10.99	3.60
Feather meal	9.28	12.70	0.0	0.0	0.0	0.00	0.0	0.0	7.85	8.92	4.09
Fish meal	1.04	4.05	0.0	0.0	0.0	0.00	0.0	0.0	6.44	10.48	3.61
Flaxseed	0.99	3.55	19.1	30.4	57.0	6.42	2.4	3.9	33.41	34.41	4.08
Flaxseed meal	1.71	4.35	16.9	30.4	49.0	6.23	2.1	5.9	3.08	3.20	3.22
Fruit and vegetable byproduct, wet	1.18	1.71	22.4	28.7	89.0	4.66	10.1	31.2	6.13	7.13	3.03
Grain screenings, source unknwn	0.92	2.40	17.7	34.1	35.9	3.65	21.8	5.0	2.91	3.91	3.01
Meat and bone meal, porcine	2.84	11.83	0.0	0.0	0.0	0.00	0.0	0.0	7.45	11.90	3.21
Molasses	0.00	0.00	0.2	0.6	50.0	0.00	0.8	60.0	0.00	0.61	3.07

Table A2.2a: Typical analyses of concentrate ingredients commonly included in western Canadian dairy cattle diets.

Feed Name	ADIP % of DM	NDIP % of DM	ADF % of DM	NDF % of DM	IVdNDF48 % of NDF	Lignin % of DM	Starch % of DM	WS Carb % of DM	Fatty Acids % of DM	Crude Fat % of DM	DE Base Mcal/kg
Oat grain	0.77	1.21	14.5	28.6	35.9	3.20	44.7	2.9	4.80	5.68	3.27
Oat hulls	0.67	4.16	39.6	73.6	17.0	6.54	10.6	1.0	1.82	1.94	2.23
Peas	1.02	3.65	7.9	12.2	71.6	0.95	43.0	8.3	1.14	2.08	3.66
Potato byproduct	0.94	1.63	10.7	14.4	80.0	3.00	57.5	7.1	1.78	2.78	3.15
Poultry byproduct meal	3.89	30.34	0.0	0.0	0.0	0.00	0.0	0.0	11.78	12.78	4.25
Rye grain	0.55	1.50	5.4	16.0	50.0	1.55	57.6	8.0	1.45	2.15	3.41
Safflower meal	1.52	2.10	40.1	55.4	25.5	13.79	1.2	5.1	3.88	5.35	2.48
Soybean hulls	1.15	3.65	47.9	66.7	88.1	2.57	1.0	2.7	1.61	1.89	2.74
Soybean meal, expeller	0.91	3.61	10.1	19.6	85.0	2.06	1.8	11.8	6.12	7.12	3.88
Soybean meal, extruded	1.41	2.12	10.6	18.4	85.0	2.15	1.5	9.3	15.08	20.42	4.14
Soybean meal, solvent 48CP	0.63	1.01	7.2	11.1	85.7	1.08	1.9	13.0	1.08	1.82	3.97
Soybeans, whole raw	0.66	1.66	7.0	11.9	84.0	1.52	4.2	9.0	16.99	20.73	4.31
Soybeans, whole roasted	0.97	2.79	10.1	18.4	84.3	1.80	1.5	9.8	15.35	21.26	4.14
Sunflower meal	1.64	2.36	29.0	40.2	36.5	9.07	1.1	9.2	1.02	2.20	2.98
Sunflower seed	0.84	1.71	24.4	35.2	15.0	7.21	0.6	4.0	37.20	39.00	4.15
Sweet corn cannery waste	6.78	10.70	32.0	56.3	72.0	3.19	10.2	4.8	3.81	5.08	2.81
Tallow	0.00	0.00	0.0	0.0		0.00	0.0		88.00	99.80	6.01
Tomato pomace	3.80	8.00	47.6	60.0	70.0	13.30	1.2	13.0	12.30	13.30	2.62
Triticale grain	0.38	1.89	4.4	14.1	41.0	1.79	61.2	8.0	1.55	1.73	3.44
Vegetable oils	0.00	0.00	0.0	0.0		0.00	0.0		88.00	100.00	6.42
Wheat bran	0.61	3.10	13.8	40.1	43.3	4.15	20.8	8.0	4.02	4.39	3.05
Wheat grain	0.45	1.59	4.2	12.5	55.7	1.52	63.0	5.9	1.78	1.98	3.56
Wheat middlings	0.69	2.77	13.2	38.7	48.6	3.77	22.9	7.9	3.85	4.35	3.05
Whey, dry	0.22	0.53	0.3	0.6	0.0	0.23	1.4	56.1	5.27	6.27	3.63
Whey, wet	0.34	0.67	0.9	2.2	0.0	0.32	1.4	50.6	0.97	1.97	3.16

Table A2.2b: Typical analyses of concentrate ingredients commonly included in western Canadian dairy cattle diets.

Feed Name	Ca	P	Mg	K	Na	Cl	S	Cu	Fe	Mn	Zn	Mo	Se
				% of DM						mg/kg			µg/kg
Bakery byproduct	0.37	0.35	0.15	0.47	0.60	0.62	0.17	6.17	237	30	31	0.00	200
Bakery byproduct, bread waste	0.22	0.26	0.09	0.33	0.65	0.93	0.21	4.19	125	21	25	1.05	200
Bakery byproduct, cereal	0.22	0.28	0.08	0.31	0.59	0.79	0.12	3.33	189	22	55	1.00	200
Bakery byproduct, cookies	0.24	0.42	0.17	0.57	0.59	0.80	0.17	6.57	172	40	41	1.08	200
Barley grain, dry, ground	0.12	0.38	0.14	0.59	0.02	0.15	0.14	5.53	89	20	34	1.05	110
Barley grain, steam-rolled	0.12	0.38	0.14	0.59	0.02	0.15	0.14	5.53	89	20	34	1.05	110
Barley malt sprouts	0.19	0.61	0.18	1.11	0.05	0.36	0.34	9.93	180	46	65	1.36	600
Beet pulp, dry	0.77	0.08	0.26	0.49	0.11	0.08	0.20	8.10	588	78	21	1.00	140
Beet pulp, dry, molasses added	1.18	0.09	0.25	0.60	0.14	0.09	0.30	8.56	614	63	24	1.04	140
Beet pulp, wet	1.13	0.13	0.26	0.57	0.10	0.11	0.24	9.40	699	60	26	1.00	140
Blood meal, high dRUP	0.14	0.28	0.05	0.43	0.42	0.35	0.74	6.05	2,267	4	33		600
Blood meal, low dRUP	0.14	0.28	0.05	0.43	0.42	0.35	0.74	6.05	2,267	4	33	0.00	700
Brewers grains, dry	0.30	0.64	0.23	0.23	0.02	0.09	0.30	15.98	350	54	86	1.92	1060
Brewers grains, wet	0.36	0.69	0.23	0.12	0.03	0.06	0.32	10.38	223	53	94	2.34	1060
Brewers yeast, dry	0.12	1.19	0.21	1.38	0.08	0.20	0.86	101.14	135	26	60	1.25	
Brewers yeast, wet	0.37	1.49	0.21	1.76	0.08	0.79	0.47	19.25	100	7	60	3.32	
Calcium fatty acid soaps	0.00	0.00	0.00	0.00	0.00	0.00	0.00	0.00	0	0	0	0.00	
Candy (not chocolate) byproduct	0.06	0.03	0.04	0.09	0.15	0.16	0.04	1.46	37	4	8	0.00	
Candy byproduct, high protein	0.32	0.49	0.30	1.22	0.23	0.21	0.23	16.03	269	57	53	1.00	
Canola meal	0.79	1.15	0.62	1.36	0.08	0.10	0.77	5.78	253	73	64	1.12	1010
Canola seed, ground	0.44	0.69	0.34	0.85	0.01	0.08	0.43	3.94	298	47	40	1.40	560
Chocolate byproduct	0.15	0.29	0.12	0.51	0.16	0.21	0.10	7.65	90	20	23	1.00	
Corn germ	0.03	1.17	0.44	1.22	0.01	0.12	0.17	5.63	99	16	72	1.00	120
Corn germ meal	0.04	0.83	0.25	0.46	0.04	0.06	0.33	6.46	135	16	79	1.06	200
Corn gluten feed, dry	0.07	1.07	0.43	1.47	0.33	0.29	0.50	5.57	146	21	74	1.24	190
Corn gluten feed, wet	0.10	1.07	0.45	1.57	0.20	0.25	0.50	6.54	179	23	77	1.30	190
Corn gluten meal	0.04	0.49	0.07	0.22	0.05	0.08	0.97	5.23	122	6	29	1.13	340
Corn grain dry, coarse grind	0.04	0.31	0.13	0.56	0.02	0.10	0.10	2.07	39	7	23	0.91	70
Corn grain dry, fine grind	0.04	0.31	0.13	0.56	0.02	0.10	0.10	2.07	39	7	23	0.91	70
Corn grain dry, medium grind	0.04	0.31	0.13	0.56	0.02	0.10	0.10	2.07	39	7	23	0.91	70
Corn grain HM, coarse grind	0.04	0.31	0.13	0.45	0.02	0.12	0.11	1.60	38	6	23	1.00	70
Corn grain HM, fine grind	0.04	0.31	0.13	0.45	0.02	0.12	0.11	1.60	38	6	23	1.00	70
Corn grain, steam-flaked	0.06	0.24	0.10	0.40	0.01	0.08	0.09	2.08	32	5	17	1.00	70
Cottonseed meal	0.25	1.31	0.70	1.74	0.17	0.08	0.48	12.48	208	23	66	1.76	300
Cottonseed, whole	0.17	0.62	0.38	1.18	0.02	0.08	0.24	7.56	72	17	36	1.14	140
Distillers grains with solubles	0.12	0.88	0.34	1.26	0.21	0.19	0.67	4.15	94	18	64	1.11	400
Distillers grains, high protein	0.08	0.64	0.23	0.75	0.21	0.20	0.64	6.65	99	21	53	1.39	400
Distillers grains, low fat	0.11	0.89	0.34	1.21	0.24	0.22	0.71	5.63	102	19	70	1.40	400
Distillers grains, modified	0.21	0.86	0.35	1.45	0.27	0.23	0.63	6.71	121	19	72	0.00	400
Distillers grains, wet	0.13	0.76	0.28	1.10	0.15	0.13	0.67	4.70	110	15	60	0.00	400
Distillers solubles	0.13	1.82	0.77	2.78	0.65	0.50	1.15	9.26	148	32	108	0.00	400
Feather meal	0.50	0.32	0.04	0.18	0.16	0.23	1.78	9.12	347	12	87	1.00	700
Fish meal	5.56	3.16	0.24	0.93	0.82	1.06	0.88	6.02	804	44	92	1.48	1500
Flaxseed	0.24	0.59	0.37	0.79	0.05	0.08	0.24	12.64	98	30	45	1.00	
Flaxseed meal	0.44	0.95	0.65	1.30	0.13	0.07	0.40	21.01	312	52	75	1.00	
Fruit and vegetable byproduct, wet	0.72	0.34	0.19	2.01	0.23	0.47	0.21	10.98	621	35	35	1.76	
Grain screenings, source unknwn	0.26	0.66	0.31	1.05	0.03	0.23	0.20	10.56	251	105	66	0.00	
Meat and bone meal, porcine	9.29	4.59	0.28	0.91	0.74	0.49	0.51	20.01	451	24	160	0.00	
Molasses	0.73	0.26	0.28	4.49	1.51	2.02	0.64	36.71	216	74	112	0.00	130

Table A2.3a: Typical mineral analyses of concentrate ingredients commonly included in western Canadian dairy cattle diets.

Feed Name	Ca	P	Mg	K	Na	Cl	S	Cu	Fe	Mn	Zn	Mo	Se
				% of DM						mg/kg			µg/kg
Oat hulls	0.16	0.15	0.11	0.56	0.02	0.13	0.09	7.63	198	49	22	1.00	
Peas	0.11	0.43	0.14	1.10	0.01	0.12	0.19	8.47	119	17	38	3.31	
Potato byproduct	0.17	0.25	0.11	1.40	0.13	0.29	0.15	6.93	348	17	22	0.00	
Poultry byproduct meal	4.31	2.48	0.16	0.89	0.38	0.55	0.74	14.14	233	45	122	1.20	
Rye grain	0.25	0.38	0.18	0.97	0.01	0.05	0.14	5.00	52	40	38	0.00	
Safflower meal	0.32	0.65	0.34	1.03	0.03	0.23	0.25	21.67	240	29	66	1.00	150
Soybean hulls	0.64	0.13	0.28	1.40	0.01	0.04	0.12	7.58	464	21	47	1.05	200
Soybean meal, expeller	0.34	0.72	0.31	2.24	0.01	0.06	0.40	15.12	196	39	53	2.72	200
Soybean meal, extruded	0.27	0.63	0.25	1.90	0.01	0.07	0.33	13.20	119	29	44	2.85	200
Soybean meal, solvent 48CP	0.40	0.74	0.33	2.42	0.02	0.06	0.41	16.06	187	41	53	4.32	200
Soybeans, whole raw	0.27	0.65	0.27	2.03	0.01	0.04	0.34	11.87	103	26	51	0.00	300
Soybeans, whole roasted	0.28	0.63	0.26	1.90	0.02	0.07	0.32	13.66	131	31	46	3.00	300
Sunflower meal	0.46	1.13	0.60	1.62	0.04	0.15	0.45	32.87	275	48	84	1.14	500
Sunflower seed	0.20	0.73	0.39	0.98	0.01	0.10	0.26	20.11	104	30	56	1.00	
Sweet corn cannery waste	0.24	0.27	0.20	1.11	0.02	0.32	0.13	8.85	470	33	40	1.51	
Tallow	0.00	0.00	0.00	0.00	0.00	0.00	0.00	0.00	0	0	0	0.00	
Tomato pomace	0.22	0.47	0.28	0.98	0.12	1.00	0.15	11.00	541	11	54	1.80	
Triticale grain	0.09	0.35	0.13	0.51	0.01	0.12	0.14	5.19	51	43	29	1.00	
Vegetable oils	0.00	0.00	0.00	0.00	0.00	0.00	0.00	0.00	0	0	0	0.00	
Wheat bran	0.13	1.06	0.43	1.22	0.03	0.11	0.19	10.76	163	133	77	1.37	500
Wheat grain	0.10	0.36	0.14	0.47	0.01	0.13	0.15	4.45	71	43	32	1.00	300
Wheat middlings	0.14	1.22	0.45	1.23	0.02	0.10	0.20	11.51	153	133	90	1.64	500
Whey, dry	0.92	0.89	0.15	2.60	0.74	1.65	0.24	1.24	11	1	5	0.00	60
Whey, wet	1.26	1.29	0.23	4.01	1.43	3.00	0.20	5.14	74	3	39	0.00	60

Table A2.3b: Typical mineral analyses of concentrate ingredients commonly included in western Canadian dairy cattle diets.

Feed Name	CP % of DM	Arginine	Histidine	Isoleucine	Leucine	Lysine	Methionine	Phenylala	Threonine	Tryptophan	Valine
		-- % of Crude Protein --									
Alfalfa meal	19.5	4.0	1.9	3.8	6.6	4.4	1.3	4.4	3.9	1.6	4.8
Apple pomace or byproduct	6.4	4.5	1.9	3.1	5.6	3.9	1.4	3.3	3.0	0.9	4.1
Bakery byproduct	12.8	4.6	2.2	3.4	6.9	2.7	1.5	4.5	3.1	1.2	4.4
Bakery byproduct, bread waste	14.9	4.7	2.6	4.0	7.8	2.9	1.7	5.4	3.4	1.2	4.4
Bakery byproduct, cereal	9.2	6.8	2.8	3.2	6.2	4.1	1.6	4.0	3.3	1.4	4.5
Bakery byproduct, cookies	12.9	4.2	1.8	3.1	7.1	1.7	1.8	4.8	3.1	0.9	4.6
Barley grain	11.8	4.9	2.2	3.4	6.8	3.6	1.7	5.1	3.3	1.2	4.8
Barley hay	10.8	2.2	1.9	5.5	6.7	3.6	1.9	4.7	4.1	1.4	4.1
Barley malt sprouts	23.9	4.6	1.9	3.2	5.7	4.7	1.5	3.4	3.4	1.3	4.6
Barley silage	10.9	1.0	1.2	3.5	4.9	2.4	1.2	3.4	2.5	1.4	4.8
Beet pulp	9.9	3.7	3.1	3.5	5.9	5.8	1.6	3.6	4.5	1.1	5.5
Blood meal	95.1	4.2	6.0	1.1	12.4	8.8	1.2	6.8	4.6	1.6	8.3
Brewers grains	28.1	5.8	2.3	4.0	8.3	3.6	2.1	5.5	3.6	1.3	5.5
Brewers yeast	50.7	4.4	2.1	4.3		6.1	1.5	3.5	4.3	1.0	4.7
Candy, not chocolate, byproduct	2.4	2.3		3.6	6.5	2.3	1.6	3.8	3.8	0.7	5.8
Candy byproduct, high protein	14.6	2.3	1.6	3.6	6.5	2.3	1.6	3.8	3.8	0.7	5.8
Canola meal	41.5	5.9	2.7	3.9	6.9	5.5	2.0	4.0	4.4	1.3	5.1
Canola seed	23.4	5.9	2.7	3.9	6.9	5.5	2.0	4.0	4.4	1.3	5.1
Casein	92.7	3.7	2.8	5.5	8.3	7.4	2.5	4.5	4.4	1.1	6.5
Chocolate byproduct	10.0	2.3	1.6	3.6	6.5	2.3	1.6	3.8	3.8	0.7	5.8
Cool season grass hay	9.2	4.1	1.9	4.0	7.4	4.9	1.6	4.8	4.1	2.1	5.2
Cool season grass silage	13.4	3.1	1.7	3.6	6.1	3.3	1.2	4.4	3.3	1.1	4.9
Corn gluten feed	23.2	4.6	2.9	3.0	8.5	3.1	1.6	3.5	3.6	0.5	4.7
Corn gluten meal	68.5	3.1	2.0	4.0	16.3	1.6	2.4	6.2	3.3	0.5	4.5
Corn grain	8.5	4.8	2.9	3.4	12.0	3.0	2.0	4.9	3.6	0.8	4.6
Corn grain HM	8.5	3.9	2.5	3.4	11.6	2.6	2.1	4.6	3.7	1.0	4.9
Corn silage	7.5	2.3	1.7	3.4	8.5	2.8	1.6	3.9	3.4	0.7	4.5
Cottonseed meal	46.7	11.6	2.7	3.0	5.5	4.0	1.4	5.3	3.1	1.2	4.2
Cottonseed, whole	23.3	10.8	2.8	3.2	5.8	4.3	1.5	5.2	3.2	1.2	4.3
Distillers grains w solubles	30.2	4.3	2.7	3.7	11.7	2.8	2.0	4.9	3.7	0.8	4.9
Feather meal	90.6	6.6	1.2	4.6	8.1	2.6	0.7	4.8	4.5	0.8	7.0
Fish meal	69.2	5.6	2.3	3.9	6.7	6.8	2.5	3.7	3.9	1.0	4.6
Flaxseed	22.8	9.1	2.2	4.1	5.9	4.1	1.8	4.6	3.6	1.5	4.9
Flaxseed meal	38.5	9.1	2.2	4.1	5.9	4.1	1.8	4.6	3.6	1.5	4.9
Grass legume mixt, grass slg	14.3	3.5	1.7	3.8	6.2	3.9	1.3	4.3	3.6	1.0	5.0
Grass legume mixt, grass hay	15.6	4.5	1.8	3.8	6.8	4.3	1.4	4.3	4.0	1.4	4.9
Grass legume mixt, legume hay	17.4	4.5	1.8	3.8	6.8	4.3	1.4	4.3	4.0	1.4	4.9
Grass legume mixt, legume slg	20.0	3.5	1.7	3.8	6.2	3.9	1.3	4.3	3.6	1.0	5.0
Grass legume mixt, mix hay	12.1	4.5	1.8	3.8	6.8	4.3	1.4	4.3	4.0	1.4	4.9
Grass legume mixt, mix silage	17.7	3.5	1.7	3.8	6.2	3.9	1.3	4.3	3.6	1.0	5.0
Legume hay	21.5	4.2	1.9	3.9	6.7	4.8	1.3	4.6	4.0	1.4	5.0
Legume silage	22.1	1.8	1.9	4.1	6.7	4.7	1.3	4.4	3.8	1.2	5.1
Meat and bone meal, porcine	56.6	7.0	1.6	2.6	5.4	4.6	1.3	3.1	2.9	0.5	3.9

Table A2.4a: Typical amino acid analyses of ingredients commonly included in western Canadian dairy cattle diets.

Feed Name	CP % of DM	Arginine	Histidine	Isoleucine	Leucine	Lysine	Methionine	Phenylala	Threonine	Tryptophan	Valine
		--- % of Crude Protein ---										
Millet hay	10.7	4.1	1.9	4.0	7.4	4.9	1.6	4.8	4.1	2.1	5.2
Millet silage	13.0	3.1	1.7	3.6	6.1	3.3	1.2	4.4	3.3	1.1	4.9
Molasses	9.3	4.9	1.6	4.4	3.6	1.0	0.2	2.7	1.6	0.5	3.4
Oat grain	12.2	6.5	1.9	3.7	7.2	3.9	1.7	4.8	3.5	1.5	5.3
Oat hay	8.5	2.2	1.9	5.5	6.7	3.6	1.9	4.7	4.1	1.4	4.1
Oat hulls	5.0	6.7	2.2	3.6	7.3	4.1	1.7	5.0	3.4	1.3	5.0
Oat silage	18.5	2.2	1.9	5.5	6.7	3.6	1.9	4.7	4.1	1.4	4.1
Pasture, grass	25.1	4.1	1.9	4.0	7.4	4.9	1.6	4.8	4.1	2.1	5.2
Pasture, grass-legume mixture	25.1	4.2	1.7	3.6	6.5	3.9	1.4	4.1	3.8	1.3	4.7
Pasture, legume	26.1	1.8	1.9	4.1	6.7	4.7	1.3	4.4	3.8	1.2	5.1
Pea hay	15.9	3.9	1.7	3.7	6.0	4.5	1.4	4.2	3.8	0.9	5.0
Pea silage	17.0	3.9	1.7	3.7	6.0	4.5	1.4	4.2	3.8	0.9	5.0
Peas	24.3	8.7	2.4	4.1	7.2	7.2	0.9	4.8	3.7	0.9	4.6
Potato byproduct	10.0	2.5	1.8	3.1	5.3	4.2	1.0	3.6	3.1	0.7	4.4
Poultry byproduct meal	65.6	4.3	1.4	2.5	4.5	3.9	1.3	2.5	2.5	0.7	3.1
Rye annual hay	7.6	4.1	1.9	4.0	7.4	4.9	1.6	4.8	4.1	2.1	5.2
Rye annual silage	16.4	3.1	1.7	3.6	6.1	3.3	1.2	4.4	3.3	1.1	4.9
Rye grain	11.8	5.0	2.3	3.2	6.2	3.6	1.6	4.4	3.3	1.1	4.6
Safflower meal	26.2	8.3	2.5	3.5	6.2	3.1	1.5	4.4	3.1	0.9	5.0
Skim milk, powder	37.4	3.8	2.9	6.2	10.6	8.8	3.0	5.3	4.6	1.7	6.9
Soybean meal, extruded	40.4	7.3	2.6	4.5	7.6	6.1	1.3	5.1	3.9	1.3	4.7
Soybean meal, solvent 48CP	52.6	7.3	2.6	4.5	7.6	6.2	1.4	5.0	4.0	1.4	4.8
Soybeans	40.0	7.3	2.6	4.5	7.6	6.1	1.3	5.0	3.9	1.3	4.7
Sunflower meal	37.0	8.0	2.4	4.0	6.2	3.5	2.2	4.5	3.6	1.3	4.8
Sunflower seed	20.1	8.0	2.5	3.9	6.1	3.7	2.1	4.5	3.6	1.4	4.8
Sunflower silage	13.3	3.9	1.7	3.7	6.0	4.5	1.4	4.2	3.8	0.9	5.0
Sweet corn cannery waste	9.8	2.3	1.7	3.4	8.5	2.8	1.6	3.9	3.4	0.7	4.5
Triticale grain	12.1	4.9	2.3	3.2	6.4	3.2	1.7	4.6	3.1	1.1	4.3
Triticale hay	10.3	3.8	2.5	3.0	5.9	1.8	1.3	4.8	2.1	1.0	3.7
Triticale silage	14.2	3.8	2.5	3.0	5.9	1.8	1.3	4.8	2.1	1.0	3.7
Wheat bran	17.4	6.9	2.8	3.2	6.2	4.1	1.5	3.9	3.2	1.8	4.8
Wheat grain	13.5	4.8	2.2	3.4	6.5	2.8	1.5	4.4	2.9	1.4	4.4
Wheat hay	9.9	2.0	3.6	4.0	6.6	4.2	1.8	4.2	4.2	1.0	5.8
Wheat middlings	19.1	6.6	2.6	3.1	6.1	4.0	1.5	4.0	3.2	1.4	4.5
Wheat silage	10.7	2.0	3.6	4.0	6.6	4.2	1.8	4.2	4.2	1.0	5.8
Wheat straw	4.5	2.0	3.6	4.0	6.6	4.2	1.8	4.2	4.2	1.0	5.8
Whey protein concentrate	37.1	2.8	1.8	5.1	8.8	6.9	1.4	3.2	5.5	1.5	4.8
Whey, dry	17.8	2.8	1.8	5.1	8.8	6.9	1.4	3.2	5.5	1.5	4.8
Whey, wet	7.4	2.2	1.7	5.4	9.0	7.2	1.4	3.0	6.2	1.6	5.1
Whole milk	25.4	3.8	2.9	6.2	10.6	8.8	3.0	5.3	4.6	1.7	6.9

Table A2.4b: Typical amino acid analyses of ingredients commonly included in western Canadian dairy cattle diets.

Feed Name	Crude Fat	Fatty Acids	C12:0	C14:0	C16:0	C16:1	C18:0	C18:1 trans	C18:1 cis	C18:2	C18:3	Other FA
	---- % of DM ----		-- % of Fatty Acids --									
Alfalfa meal	2.35	1.61	1.17	0.58	20.10	1.71	3.35		3.02	21.07	44.93	4.09
Almond hulls	2.52	1.26	0.22	0.05	14.49	0.26	3.91		50.82	23.39	3.25	3.61
Apple pomace or byproduct	5.97	1.88	0.60	1.20	26.90	0.60	3.90		7.80	48.80	10.10	0.10
Bakery byproduct	8.68	7.68	0.49	3.16	15.82	0.18	9.29	7.77	26.41	33.51	0.85	2.53
Bakery byproduct, bread waste	5.76	4.76		5.21	18.86		1.02		13.90	56.83	3.81	0.37
Bakery byproduct, cereal	2.95	1.95		5.21	18.86		1.02		13.90	56.83	3.81	0.37
Bakery byproduct, cookies	10.04	9.04	0.49	3.16	15.82	0.18	9.29	7.77	26.41	33.51	0.85	2.53
Barley grain	2.31	1.31		0.32	22.97	0.06	1.53		13.54	55.93	4.34	1.32
Barley hay	2.74	1.41	0.80	5.50	43.40		4.10		7.30	12.30	2.40	24.20
Barley malt sprouts	2.73	1.46		0.32	22.97	0.06	1.53		13.54	55.93	4.34	1.32
Barley silage	3.08	2.06	0.80	5.50	43.40		4.10		7.30	12.30	2.40	24.20
Beet pulp	1.06	0.63			26.66	0.39	0.89	0.14	11.55	49.83	6.35	4.19
Blood meal	1.69	1.31	0.11	1.48	21.62	1.02	21.66	5.13	26.47	14.89	0.46	7.16
Brewers grains, dry	9.03	8.31	0.14	0.73	26.69	0.20	2.24		14.61	48.87	4.57	1.95
Brewers grains, wet	9.52	7.61	0.03	0.40	24.49	0.20	1.84		11.23	53.82	5.37	2.63
Calcium soaps	84.50	84.50	0.20	1.60	50.80		4.10		35.70	7.00	0.20	0.40
Candy, not chocolate, byproduct	1.25	0.25	0.23	0.35	18.55	0.23	7.94	17.44	38.75	15.56	0.20	0.74
Candy byproduct, high protein	12.11	11.11	6.54	5.13	18.24	0.14	18.81	9.08	30.81	7.78	0.58	2.89
Canola meal	3.51	2.51		0.19	9.76	0.91	2.24	0.61	45.21	31.45	7.71	1.92
Canola seed	40.46	39.46		0.19	9.76	0.91	2.24	0.61	45.21	31.45	7.71	1.92
Chocolate byproduct	21.68	20.68	6.54	5.13	18.24	0.14	18.81	9.08	30.81	7.78	0.58	2.89
Cool season grass hay	2.35	0.96	0.89		15.22	1.48	1.29		2.52	16.62	55.50	6.49
Cool season grass silage	3.63	1.84	6.56	0.54	16.76	1.67	1.94		3.80	19.96	44.30	4.46
Corn gluten feed, dry	3.91	3.38	0.03	0.99	23.94	0.20	2.51		17.19	50.49	3.21	1.44
Corn gluten feed, wet	3.81	3.09	0.31	0.06	20.87	0.32	4.30	0.13	19.08	49.97	2.90	2.07
Corn gluten meal	2.44	1.44		0.22	13.62	0.08	2.17		22.22	57.46	2.77	1.47
Corn grain dry	3.85	3.84		2.33	13.21	0.12	1.99		24.09	55.70	1.62	0.93
Corn grain HM, coarse grind	3.58	3.57		2.33	13.21	0.12	1.99		24.09	55.70	1.62	0.93
Corn grain HM, fine grind	3.58	3.57		0.26	13.57	0.19	1.83		25.99	55.08	1.64	1.44
Corn grain, steam-flaked	3.14	3.14		0.87	12.92	0.08	1.86		23.17	58.38	1.82	0.90
Corn silage	2.96	2.32	0.31	0.46	17.83	0.36	2.42	0.00	19.24	47.74	8.25	3.40
Cottonseed meal	3.60	3.06		0.94	25.80	0.52	2.95	0.05	18.33	50.20	0.30	0.92
Cottonseed, whole	18.62	18.26		0.69	23.91	0.55	2.33		15.24	56.48	0.19	0.61
Distillers grains w solubles, dry	12.55	11.39	0.12	0.14	14.05	0.13	2.39	0.01	24.57	56.11	1.68	0.81
Distillers grains w solubles, wet	9.31	8.31	0.30	0.25	15.00	0.10	2.50	0.05	18.00	55.00	8.00	0.80
Distillers solubles	10.99	9.99	0.30	0.25	15.00	0.10	2.50	0.05	18.00	55.00	8.00	0.80
Fat, canola oil	100.00	88.00	0.10	0.10	4.36	0.28	2.05	3.53	57.28	18.99	7.64	5.67
Fat, corn oil	100.00	88.00			11.08		1.55		26.95	58.95	1.10	0.38
Fat, cottonseed oil	100.00	88.00		0.83	25.97	0.57	3.00		20.16	48.93	0.10	0.44
Fat, flaxseed oil	100.00	88.00		0.16	5.74	0.18	4.30		18.88	14.15	55.95	0.65
Fat, lard	100.00	88.00	0.20	1.30	23.80	2.70	13.50		41.20	10.20	1.00	6.10
Fat, safflower oil	100.00	88.00		0.10	10.77		11.97		14.33	60.63	0.30	1.90
Fat, soybean oil	100.00	88.00	0.11	0.11	10.83	0.14	3.89		22.82	53.75	8.23	0.13
Fat, sunflower oil	100.00	88.00			7.33	0.09	10.65	0.59	43.39	35.49	0.79	1.67
Fat, tallow	99.80	88.00	0.09	3.00	24.43	3.79	17.92	3.99	41.62	1.09	0.53	3.54
Feather meal	8.92	7.85	0.34	1.09	24.33	6.51	8.27	1.09	32.51	13.19	0.54	12.13
Fish meal	10.48	6.44		10.35	28.46	13.01	6.00	0.20	10.97	1.10	0.96	28.96
Flaxseed	34.41	33.41		0.16	5.74	0.18	4.30		18.88	14.15	55.95	0.64
Flaxseed meal	3.20	3.08		0.16	5.74	0.18	4.30		18.88	14.15	55.95	0.64

Table A2.5a: Typical fatty acid analyses of ingredients commonly included in western Canadian dairy cattle diets.

Feed Name	Crude Fat	Fatty Acids	C12:0	C14:0	C16:0	C16:1	C18:0	C18:1 trans	C18:1 cis	C18:2	C18:3	Other FA
	----- % of DM -----							--------------------------- % of Fatty Acids ---------------------------				
Grass legume mixt, grass slg	3.75	1.98	9.27	0.60	17.79	1.79	2.65		2.93	17.94	41.51	5.55
Grass legume mixt hay	3.39	1.93	1.15	0.62	18.98	1.77	2.79	0.17	1.85	33.21	33.39	6.08
Grass legume mixt, legume slg	3.81	1.99	9.27	0.60	17.79	1.79	2.65		2.93	17.94	41.51	5.55
Grass legume mixt, mix silage	4.04	2.03	9.27	0.60	17.79	1.79	2.65		2.93	17.94	41.51	5.55
Legume hay	2.55	1.54	1.36	0.85	25.01	2.23	4.01	0.35	2.43	18.49	36.79	8.47
Legume silage	3.22	1.98	11.98	0.66	18.81	1.91	3.35		2.05	15.91	38.71	6.63
Meat and bone meal, porcine	11.90	7.45	0.08	1.68	29.47	2.41	17.50	1.27	40.43	3.70	0.08	3.38
Millet hay	1.95	1.09	2.86	0.89	20.64	0.43	2.42		10.18	30.37	25.53	6.68
Millet silage	2.79	1.47	2.86	0.89	20.64	0.43	2.42		10.18	30.37	25.53	6.68
Molasses	0.61	0.00			17.99	0.34	3.61		12.98	54.94	7.46	2.68
Oat grain	5.68	4.80		0.87	17.65	0.16	1.32		34.78	42.01	1.85	1.38
Oat hay mid-maturity	2.36	1.45	1.19	0.43	16.44	0.48	1.33	0.06	2.53	23.38	49.90	4.26
Oat hulls	1.94	1.82			19.72		2.51	2.56	34.23	37.67	2.40	0.91
Oat silage	4.23	2.24	6.56	0.54	16.76	1.67	1.94		3.80	19.96	44.30	4.46
Pasture, grass	4.10	2.32	0.84	0.24	13.49		1.07		2.07	13.34	66.49	2.46
Pasture, grass-legume mixture	4.10	2.32	1.43	0.33	15.22	1.52	2.39		2.62	18.19	53.84	4.47
Pasture, legume	3.50	2.00	11.98	0.66	18.81	1.91	3.35		2.05	15.91	38.71	6.63
Pea hay	2.98	1.69	0.43	0.28	17.97	0.15	6.71		19.74	38.88	12.98	2.85
Pea silage	3.80	1.68	0.43	0.28	17.97	0.15	6.71		19.74	38.88	12.98	2.85
Peas	2.08	1.14		0.30	23.00	0.10	1.50		13.50	55.90	4.30	1.40
Potato byproduct	2.78	1.78	0.35	0.49	12.18	0.56	10.70	31.21	35.65	5.12	1.15	2.60
Rye annual fresh	5.03	3.07	0.84	0.24	13.49		1.07		2.07	13.34	66.49	2.46
Rye annual hay	4.46	2.53	4.60	3.30	26.20	1.70	5.40		11.00	18.40	9.40	20.00
Rye annual hay, mid-maturity	2.89	1.47	4.60	3.30	26.20	1.70	5.40		11.00	18.40	9.40	20.00
Rye annual silage	4.10	1.95	0.66	1.87	20.40	1.19	2.25		5.14	19.12	39.07	10.31
Rye grain	2.15	1.45		0.18	15.93	0.59	0.53		16.46	56.32	9.23	0.76
Safflower meal	5.35	3.88			5.40		1.60		13.20	79.50	0.30	
Soybean meal, extruded	20.42	15.08		0.07	11.55	0.09	3.71	1.42	18.13	54.77	9.52	0.76
Soybean meal, solvent 48CP	1.82	1.08		0.83	17.28		4.45	0.43	13.22	54.16	8.43	1.20
Soybeans, whole raw	20.73	16.99	0.58	0.20	11.93	0.08	4.05		21.99	52.43	7.59	1.17
Soybeans, whole roasted	21.26	15.35	0.00	0.11	11.80	0.08	4.30		23.58	52.36	6.99	0.79
Sunflower meal	2.20	1.02		0.76	11.59		4.37		41.93	38.71	0.59	2.05
Sunflower seed	39.00	37.20		0.10	5.20	0.10	4.10		39.40	47.90	0.40	2.80
Sunflower silage	5.39	3.06	0.43	0.28	17.97	0.15	6.71		19.74	38.88	12.98	2.85
Triticale grain	1.73	1.55		0.23	19.50		1.08		14.66	59.45	4.32	0.76
Triticale hay	2.29	1.46	1.12	0.42	11.42	2.23	0.92		1.49	14.53	62.82	5.06
Triticale silage	3.47	2.48	1.12	0.42	11.42	2.23	0.92		1.49	14.53	62.82	5.06
Wheat bran	4.39	4.02		0.10	17.05		1.08		17.81	59.09	4.73	0.14
Wheat grain	1.98	1.78		0.23	19.50		1.08		14.66	60.20	4.32	
Wheat hay	2.09	1.01	1.19	0.43	16.44	0.48	1.33	0.06	2.53	23.38	49.90	4.26
Wheat middlings	4.35	3.85		0.10	17.09	0.12	1.17		17.69	57.78	4.71	1.35
Wheat silage	3.06	1.53	6.56	0.54	16.76	1.67	1.94		3.80	19.96	44.30	4.46
Wheat straw	1.49	0.55	1.19	0.43	16.44	0.48	1.33	0.06	2.53	23.38	49.90	4.26
Whey	6.27	5.27	0.72	6.75	35.74	0.95	17.81	2.64	27.08	6.89		1.42

Table A2.5b: Typical fatty acid analyses of ingredients commonly included in western Canadian dairy cattle diets.

BW (kg)	Breed[a]	ADG (g/d)	DMI[b] (kg/d)	ME (Mcal/d)	NEm (Mcal/d)	MP (g/d)	CP[c] (g/d)	CP (% of DMI)
25	SB	200	0.36	1.69	0.82	83	87	24.2
		400	0.49	2.33	0.82	132	139	28.2
30	SB	200	0.40	1.88	0.94	86	91	22.7
		400	0.54	2.54	0.94	136	143	26.6
		600	0.69	3.23	0.94	186	196	28.5
35	SB	200	0.44	2.06	1.06	89	94	21.4
		400	0.58	2.74	1.06	140	147	25.2
		600	0.73	3.45	1.06	190	200	27.2
		800	0.89	4.19	1.06	240	253	28.3
40	LB	200	0.48	2.23	1.17	91	96	19.8
		400	0.64	2.93	1.17	142	149	23.4
		600	0.80	3.66	1.17	192	202	25.4
		800	0.96	4.42	1.17	242	254	26.5
45	LB	200	0.52	2.40	1.28	94	99	19.0
		400	0.68	3.11	1.28	145	152	22.5
		600	0.84	3.86	1.28	195	205	24.5
		800	1.01	4.64	1.28	245	258	25.6
50	LB	200	0.56	2.56	1.38	97	102	18.3
		400	0.71	3.29	1.38	148	155	21.8
		600	0.88	4.05	1.38	198	209	23.7
		800	1.05	4.85	1.38	249	262	24.9
		1,000	1.23	5.66	1.38	299	315	25.6

Table B1: Daily energy and protein requirements of young replacement calves fed only milk or milk replacer.

[a]SB = small breed (based on Jersey, MatBW = 530 kg) and LB = large breed (based on Holstein, MatBW = 700 kg).

[b]Dry Matter Intake necessary to meet requirement for ME when fed milk replacer containing 4.7 Mcal ME/kg of DM (SB calves) or 4.6 ME/kg of DM (LB calves). Not a prediction of actual DMI.

[c]Assumes all milk protein with MP/CP of 0.95.

BW (kg)	ADG (g/d)	Diet	DMI[a] (kg/d)	ME (Mcal/d)	NEm (Mcal/d)	MP (g/d)	CP[b] (g/d)	CP (% of DMI)
30	200	80:20[c]	0.43	1.96	0.93	86	94	21.8
	200	40:60[d]	0.53	2.02	0.93	91	109	20.6
	400	80:20	0.58	2.62	0.93	135	148	25.6
	400	40:60	0.72	2.75	0.93	142	170	23.6
	600	80:20	0.73	3.32	0.93	184	202	27.6
	600	40:60	0.93	3.52	0.93	192	232	25.0
40	200	80:20	0.51	2.33	1.16	92	101	19.7
	200	40:60	0.63	2.39	1.16	98	118	18.7
	400	80:20	0.67	3.03	1.16	142	156	23.4
	400	40:60	0.83	3.16	1.16	150	180	21.7
	600	80:20	0.83	3.77	1.16	192	211	25.4
	600	40:60	1.05	3.98	1.16	202	243	23.2
	800	80:20	1.00	4.54	1.16	241	265	26.5
	800	40:60	1.27	4.82	1.16	253	305	24.0
50	400	80:20	0.75	3.41	1.37	149	163	21.8
	400	40:60	0.93	3.54	1.37	157	189	20.3
	600	80:20	0.92	4.18	1.37	199	219	23.8
	600	40:60	1.16	4.40	1.37	210	253	21.9
	800	80:20	1.10	4.98	1.37	250	274	25.0
	800	40:60	1.39	5.28	1.37	263	317	22.8
	1,000	80:20	1.28	5.80	1.37	300	330	25.8
	1,000	40:60	1.63	6.19	1.37	316	380	23.4
60	400	80:20	0.83	3.75	1.57	155	170	20.6
	400	40:60	1.03	3.90	1.57	164	198	19.3
	600	80:20	1.00	4.56	1.57	206	227	22.6
	600	40:60	1.26	4.79	1.57	218	263	20.9
	800	80:20	1.19	5.39	1.57	258	283	23.8
	800	40:60	1.50	5.70	1.57	272	328	21.8
	1,000	80:20	1.38	6.25	1.57	309	340	24.7
	1,000	40:60	1.75	6.64	1.57	326	392	22.4

Table B2: Daily energy and protein requirements of small-breed calves fed milk or milk replacer and starter at two different ratios expressed as the proportion of DM from milk or milk replacer to the proportion of DM from starter.
[a]Total Dry Matter Intake with mean ME density needed to meet ME requirements. Not a prediction of actual DMI.
[b]Total dietary CP needed, assuming all-milk protein milk replacer.
[c]Assumes MR contains 4.9 Mcal ME/kg DM and starter contains 3.1 Mcal ME/kg DM.
[d]Assumes MR contains 4.7 Mcal ME/kg DM and starter contains 3.2 Mcal ME/kg DM.

BW (kg)	ADG (g/d)	Diet	DMI[a] (kg/d)	ME (Mcal/d)	NEm (Mcal/d)	MP (g/d)	CP[b] (g/d)	CP (% of DMI)
50	400	80:20[c]	0.79	3.40	1.37	147	162	20.6
	400	40:60[d]	0.93	3.54	1.37	155	187	20.0
	600	80:20	0.97	4.18	1.37	197	217	22.4
	600	40:60	1.16	4.40	1.37	207	250	21.6
	800	80:20	1.15	4.98	1.37	247	271	23.6
	800	40:60	1.39	5.28	1.37	259	312	22.5
60	400	80:20	0.87	3.75	1.57	153	168	19.4
	400	40:60	1.03	3.90	1.57	162	195	19.0
	600	80:20	1.05	4.55	1.57	204	224	21.2
	600	40:60	1.26	4.79	l.57	214	258	20.5
	800	80:20	1.25	5.38	1.57	254	279	22.4
	800	40:60	l.50	5.70	1.57	267	322	21.4
	1,000	80:20	1.44	6.24	1.57	305	335	23.2
	1,000	40:60	1.75	6.64	1.57	320	386	22.0
70	400	80:20	0.94	4.08	1.76	159	174	18.4
	400	40:60	1.11	4.24	1.76	168	202	18.1
	600	80:20	1.14	4.91	1.76	210	231	20.3
	600	40:60	1.35	5.15	1.76	221	267	19.7
	800	80:20	1.33	5.77	1.76	261	287	21.5
	800	40:60	1.60	6.10	1.76	275	332	20.6
	1,000	80:20	1.54	6.65	1.76	312	343	22.3
	1,000	40:60	1.86	7.07	1.76	328	396	21.3
80	600	80:20	1.21	5.25	1.90	216	237	19.5
	600	40:60	1.45	5.50	1.90	228	275	19.0
	800	80:20	1.42	6.13	1.90	268	294	20.7
	800	40:60	1.70	6.47	1.90	282	340	20.0
	1,000	80:20	1.63	7.03	1.90	320	351	21.6
	1,000	40:60	1.96	7.47	1.90	337	406	20.6

Table B3: Daily energy and protein requirements of large-breed calves fed milk or milk replacer and starter at two different ratios expressed as the proportion of DM from milk or milk replacer to the proportion of DM from starter.
[a]*Total Dry Matter Intake with mean ME density needed to meet ME requirements. Not a prediction of actual DMI.*
[b]*Total dietary CP needed, assuming all-milk protein milk replacer.*
[c]*Assumes milk replacer contains 4.6 Mcal ME/kg DM and starter contains 3.2 Mcal ME/kg DM.*
[d]*Assumes milk replacer contains 4.7 Mcal ME/kg DM and starter contains 3.2 Mcal ME/kg DM.*

Minerals/Vitamins[a]	Milk Replacer[b]	Starter[c]	Grower[d]
Ca, % of DM	0.80	0.75	0.65
P, % of DM	0.60	0.37	0.33
Mg, % of DM	0.15	0.15	0.16
K, % of DM	1.10	0.60	0.60
Na, % of DM	0.40	0.22	0.20
Cl, % of DM	0.32	0.17	0.15
Co, mg/kg of DM	NA	0.2	0.2
Cu, mg/kg of DM	5	12	12
I, mg/kg of DM	0.8	0.8	0.5
Fe, mg/kg of DM	85	60	55
Mn, mg/kg of DM	60	40	60
Se, mg/kg of DM	0.3	0.3	0.3
Zn, mg/kg of DM	65	55	50
Vitamin A, IU/kg of DM	11,000	3,700	3,700
Vitamin D, IU/kg of DM	3,200	1,100	1,100
Vitamin E, IU/kg of DM	200	67	67

Table B4: Recommended concentrations of minerals and fat-soluble
   vitamins in milk replacers, calf starters and calf grower diets for calves
   between 35 and 125 kg of body weight and gaining between 0.5 and 1.2
   kg/day.
[a]Microbial synthesis of vitamin K within the intestines is generally
   adequate; supplemental vitamin K is usually not required.
[b]These values assume a 60 kg calf that is consuming 0.6 kg of milk
   replacer solids. Concentrations should be reduced if calves are fed
   substantially greater amounts of milk replacer (e.g., ≥1 kg/d of solids).
[c]These values assume an 80 kg calf consuming 2.4 kg of starter DM.
[d]These values assume a 110 kg calf consuming 3.3 kg of grower DM.

Water-soluble Vitamins		Concentration Required in Milk Replacer mg/kg of DM
B1	Thiamin	6.5
B2	Riboflavin	6.5
B3	Niacin	10
B5	Pantothenic Acid	13
B6	Pyridoxine	6.5
B7	Biotin	0.1
B9	Folic Acid	0.5
B12	Cobalamin	0.07
	Choline	1,000

Table B5: Recommended concentrations of water-
   soluble vitamins in milk replacers.

BW (kg)	ADG (g/d)	DMI^a (kg/d)	ME (Mcal/d)	NEm (Mcal/d)	MP (g/d)	CP (g/d)	CP (% of DM)
55	400	1.31	3.94	1.73	168	224	17.1
	600	1.59	4.77	1.73	221	295	18.6
	800	1.87	5.62	1.73	274	366	19.5
65	400	1.44	4.32	1.97	177	235	16.4
	600	1.72	5.18	1.97	231	307	17.8
	800	2.02	6.06	1.97	285	379	18.8
75	400	1.56	4.68	2.19	184	246	15.8
	600	1.85	5.56	2.19	239	319	17.2
	800	2.16	6.48	2.19	294	393	18.2
85	600	1.98	5.93	2.40	243	324	16.4
	800	2.29	6.87	2.40	297	396	17.3
	1,000	2.61	7.83	2.40	352	469	18.0
95	600	2.10	6.29	2.61	250	334	15.9
	800	2.41	7.25	2.61	306	408	16.9
	1,000	2.74	8.23	2.61	361	482	17.6
105	600	2.21	6.63	2.82	258	344	15.6
	800	2.54	7.61	2.82	314	419	16.5
	1,000	2.87	8.61	2.82	370	494	17.2
	1,200	3.21	9.64	2.82	426	569	17.7
115	600	2.32	6.96	3.01	365	354	15.2
	800	2.65	7.96	3.01	322	429	16.2
	1,000	2.99	8.98	3.01	379	505	16.9
	1,200	3.34	10.03	3.01	436	581	17.4
125	600	2.43	7.28	3.21	272	363	15.0
	800	2.77	8.30	3.21	330	440	15.9
	1,000	3.11	9.34	3.21	388	517	16.6
	1,200	3.47	10.40	3.21	446	594	17.1
	1,400	3.83	11.48	3.21	504	671	17.5

Table B6: Daily energy and protein requirements of weaned large- or small-breed calves fed only solid feeds. ADG - average daily gain; DMI - dry matter intake; ME - metabolizable energy; NEm - net energy maintenance; MP - metabolizable protein; CP - crude protein. ^{a}Assumes starter contains 3.0 Mcal ME/kg DM.

	Growing Holstein Calves and Heifers					
Age, days	30	100	225	350	475	600
Body weight, kg	65	120	230	330	420	530
Growth Rate, kg/d	0.7	0.7	0.9	0.8	0.7	0.9
Dry matter (DM) intake, kg/d	1.4	3.9	6.6	8.5	9.8	11.0
Metabolizable energy, Mcal/kg of DM	3.68	2.26	2.09	1.95	1.92	2.12
Crude protein (CP), % of DM	21.0	16.6	14.4	12.6	11.7	12.7
Metabolizable protein, % of DM	16.5	9.5	8.1	6.8	6.1	14.0
Rumen degradable protein, % of DM	—	10.0	10.0	10.0	10.0	10.0
Rumen degradable protein, % of CP	—	60	69	79	85	79
Rumen undegradable protein, % of DM	—	6.6	4.4	2.6	1.7	2.7
Rumen undegradable protein, % of CP	—	40	31	21	15	21
Neutral detergent fibre (NDF), min % of DM	—	25-33	25-33	25-33	25-33	25-33
Forage NDF, min % of DM	—	19-25	19-25	19-25	19-25	19-25
Starch max, % of DM (varies)	—	15-20	15-20	15-20	15-20	15-20
Macrominerals, % of DM						
Ca	0.59	0.78	0.58	0.44	0.37	0.39
P	0.45	0.32	0.26	0.21	0.18	0.19
Mg	0.15	0.14	0.12	0.12	0.12	0.10
K	1.00	0.51	0.52	0.54	0.56	0.60
Na	0.35	0.17	0.16	0.16	0.15	0.16
Cl	0.28	0.14	0.14	0.13	0.13	0.13
S	—	0.20	0.20	0.20	0.20	0.20
Trace minerals, mg/kg of DM						
Cu	5	16	16	15	15	17
Co	—	0.20	0.20	0.20	0.20	0.20
I	0.78	0.69	0.58	0.54	0.53	0.54
Fe	90	61	46	32	24	28
Mn	50	49	44	40	38	43
Se	0.3	0.3	0.3	0.3	0.3	0.3
Zn	70	47	41	36	34	35
Vitamins, IU/kg of DM						
Vitamin A	5,218	3,390	3,829	4,265	4,698	5,288
Vitamin D	1,518	924	1,044	1,163	1,281	1,442
Vitamin E	86	49	56	62	68	77

Table B7: Predicted nutrient concentrations in dietary dry matter needed to meet the requirements for growing Holstein calves and heifers at various ages; mature body weight = 700 kg.

| | Lactating Holstein Cows by Parity (Body Weight) and Days in Milk | | | | |
	Primiparous (570 kg)		Multiparous (700 kg)		
Days in milk	15	150	20	100	200
Milk, kg	33	39	53	55	43
Fat, % in milk	3.9	3.6	3.7	3.5	3.8
Protein, % in milk	3.1	3.0	2.8	2.8	3.3
Dry matter (DM) intake, kg/d	20.8	23.9	25.8	29.4	27.4
Metabolizable energy, Mcal/kg of DM	2.39	2.61	2.58	2.73	2.60
Net energy lactation, Mcal/kg of DM	1.58	1.72	1.70	1.80	1.73
Crude protein (CP), % of DM	16.2	16.0	17.5	17.4	17.5
Metabolizable protein, % of DM	9.8	9.8	10.9	10.2	10.1
Rumen degradable protein, % of DM	10.0	10.0	10.0	10.0	10.0
Rumen degradable protein, % of CP	62	63	57	57	57
Rumen undegradable protein, % of DM	6.2	7.0	7.5	7.4	7.5
Rumen undegradable protein, % of CP	38	37	43	43	43
Neutral detergent fibre (NDF), min % of DM	25-33	25-33	25-33	25-33	25-33
Forage NDF, min % of DM	19-25	19-25	19-25	19-25	19-25
Starch max, % of DM (varies)	22-30	22-30	22-30	22-30	22-30
Macrominerals, % of DM					
Ca	0.57	0.57	0.64	0.60	0.58
P	0.35	0.35	0.39	0.37	0.35
Mg	0.17	0.17	0.18	0.18	0.17
K	1.03	0.97	1.10	1.00	0.99
Na	0.21	0.21	0.23	0.22	0.21
Cl	0.29	0.30	0.34	0.32	0.29
S	0.20	0.20	0.20	0.20	0.20
Trace minerals, mg/kg of DM					
Cu	9	8	10	8	10
Co	0.20	0.20	0.20	0.20	0.20
I	0.46	0.42	0.47	0.42	0.41
Fe	16	16	21	19	16
Mn	28	26	31	28	27
Se	0.3	0.3	0.3	0.3	0.3
Zn	57	58	66	62	61
Vitamins, IU/kg of DM					
Vitamin A	3,021	2,796	3,687	3,303	3,103
Vitamin D	1,099	954	1,085	952	1,021
Vitamin E	22	19	22	19	20

Table B8: Predicted nutrient concentrations in dietary dry matter needed to meet the requirements for lactating primiparous and multiparous Holstein cows. Energy and protein requirements for lactating cows have been adjusted for growth (0.19 and 0.1 kg/d) for primiparous versus multiparous cows and changes in energy reserves (-1.00, 0.20, -1.70, 0.21, and 0.21 kg/d) for the respective groups. Days pregnant were set at 10, 60, and 110 for cows at 100, 150, and 200 days in milk.

	Dry Holstein Cows	
Days before calving	60-21	less than 21
Body weight, kg	740	740
Growth Rate, kg/d	0.1	0.1
Dry matter (DM) intake, kg/d	13.9	12.3
Metabolizable energy, Mcal/kg	1.93	2.25
Net energy lactation, Mcal/kg	1.28	1.49
Crude protein (CP), % of DM	11.9	14.3
Metabolizable protein, % of DM	5.2	6.7
Rumen degradable protein, % of DM	10.0	10.0
Rumen degradable protein, % of CP	84	70
Rumen undegradable protein, % of DM	1.9	4.3
Rumen undegradable protein, % of CP	16	30
Neutral detergent fibre (NDF), min % of DM	25-33	25-33
Forage NDF, min % of DM	19-25	19-25
Starch max, % of DM (varies)	15-20	15-20
Macrominerals, % of DM		
Ca	0.31	0.39
P	0.19	0.21
Mg	0.13	0.14
K	0.62	0.69
Na	0.16	0.17
Cl	0.13	0.14
S	0.20	0.20
Trace minerals, mg/kg of DM		
Cu	18	19
Co	0.20	0.20
I	0.51	0.54
Fe	13	15
Mn	38	43
Se	0.3	0.3
Zn	30	32
Vitamins, IU/kg of DM		
Vitamin A	5,850	6,630
Vitamin D	1,595	1,810
Vitamin E	85	181

Table B9: Predicted nutrient concentrations in dietary dry matter needed to meet the requirements for dry, pregnant mature Holstein cows. Energy and protein requirements have been adjusted for changes in energy reserves (-0.36 kg/d) for dry cows at less than 21 days prepartum.

			Growing Jersey Calves and Heifers			
Age, days	30	100	225	350	475	600
Body weight, kg	45	90	175	245	310	400
Growth Rate, kg/d	0.5	0.6	0.7	0.6	0.7	0.7
Dry matter (DM) intake, kg/d	I.O	3.0	5.1	6.5	7.4	8.3
Metabolizable energy, Mcal/kg of DM	3.69	2.41	2.16	2.02	2.13	2.25
Crude protein (CP), % of DM	22.9	17.8	14.4	13.1	13.1	12.9
Metabolizable protein, % of DM	18.2	10.6	8.2	7.2	6.9	6.6
Rumen degradable protein, % of DM	—	10.0	10.0	10.0	10.0	10.0
Rumen degradable protein, % of CP		56	69	76	76	78
Rumen undegradable protein, % of DM	—	7.8	4.4	3.1	3.1	2.9
Rumen undegradable protein, % of CP		44	31	24	24	22
Neutral detergent fibre (NDF), min % of DM	—	25-33	25-33	25-33	25-33	25-33
Forage NDF, min % of DM	—	19-25	19-25	19-25	19-25	19-25
Starch max, % of DM (varies)	—	15-20	15-20	15-20	15-20	15-20
Macrominerals, % of DM						
Ca	0.75	0.84	0.58	0.44	0.43	0.39
P	0.55	0.34	0.26	0.21	0.20	0.19
Mg	0.15	0.14	0.12	0.12	0.12	0.10
K	1.20	0.50	0.52	0.53	0.56	0.60
Na	0.43	0.17	0.16	0.16	0.16	0.16
Cl	0.34	0.14	0.14	0.13	0.13	0.13
S	—	0.20	0.20	0.20	0.20	0.20
Trace minerals, mg/kg						
Cu	6	17	15	15	16	17
Co	—	0.20	0.20	0.20	0.20	0.20
I	1.08	0.77	0.64	0.61	0.60	0.61
Fe	110	68	46	32	32	29
Mn	60	52	44	39	42	44
Se	0.3	0.3	0.3	0.3	0.3	0.3
Zn	84	49	41	36	36	35
Vitamins, IU/kg						
Vitamin A	6,084	3,286	3,745	4,162	4,592	5,273
YitaminD	1,770	896	1,021	1,135	1,252	1,438
VitaminE	90	48	54	61	67	77

Table B10: Predicted nutrient concentrations in dietary dry matter needed to meet the requirements for growing Jersey calves and heifers at various ages; mature body weight = 525 kg.

| | Lactating Jersey Cows by Parity (Body Weight) and Days in Milk | | | | |
	Primiparous (425 kg)		Multiparous (520 kg)		
Days in milk	15	150	20	100	200
Milk, kg	22	27	35	37	31
Fat, % in milk	4.9	4.9	5.0	4.8	4.8
Protein, % in milk	3.9	3.7	3.5	3.5	3.7
Dry matter (DM) intake, kg/d	16.5	19.4	20.5	23.5	21.9
Metabolizable energy, Mcal/kg of DM	2.41	2.72	2.67	2.80	2.68
Net energy lactation, Mcal/kg of DM	1.59	1.79	1.76	1.85	1.78
Crude protein (CP), % of DM	16.8	17.6	18	18	17.6
Metabolizable protein, % of DM	10.2	10.0	11.1	10.5	10.2
Rumen degradable protein, % of DM	10.0	10.0	10.0	10.0	10.0
Rumen degradable protein, % of CP	60	57	56	56	57
Rumen undegradable protein, % of DM	6.8	7.6	8.0	8.0	7.6
Rumen undegradable protein, % of CP	40	43	44	44	43
Neutral detergent fibre (NDF), min % of DM	25-33	25-33	25-33	25-33	25-33
Forage NDF, min % of DM	19-25	19-25	19-25	19-25	19-25
Starch max, % of DM (varies)	22-30	22-30	22-30	22-30	22-30
Macrominerals, % of DM					
Ca	0.56	0.56	0.63	0.60	0.57
P	0.33	0.34	0.37	0.35	0.34
Mg	0.16	0.16	0.17	0.17	0.16
K	0.96	0.89	1.01	0.92	0.93
Na	0.20	0.20	0.21	0.21	0.20
Cl	0.27	0.27	0.31	0.29	0.27
S	0.20	0.20	0.20	0.20	0.20
Trace minerals, mg/kg					
Cu	9	8	9	8	8
Co	0.20	0.20	0.20	0.20	0.20
I	0.45	0.41	0.46	0.41	0.41
Fe	13	14	17	16	14
Mn	25	23	28	25	25
Se	0.3	0.3	0.3	0.3	0.3
Zn	53	54	59	56	53
Vitamins, IU/kg					
Vitamin A	2,836	2,405	2,796	2,520	2,616
Yitamin D	1,031	875	1,017	884	951
Vitamin E	21	17	20	18	19

Table B11: Predicted nutrient concentrations in dietary dry matter needed to meet the requirements for lactating primiparous and multiparous Jersey cows. Energy and protein requirements for lactating cows have been adjusted for growth (0.14 and 0.06 kg/d) for primiparous versus multiparous cows and changes in energy reserves (-0.75, 0.15, -1.28, 0.16, and 0.16 kg/d) for the respective groups. Days pregnant were set at 10, 60, and 110 for cows at 100, 150, and 200 DIM.

	Dry Jersey Cows	
Days before calving	60-21	less than 21
Body weight, kg	555	555
Growth Rate, kg/d	0.06	0.06
Dry matter (DM) intake, kg/d	10.4	9.4
Metabolizable energy, Mcal/kg	2.04	2.19
Net energy lactation, Mcal/kg	1.36	1.44
Crude protein (CP), % of DM	11.8	13.7
Metabolizable protein, % of DM	5.5	6.1
Rumen degradable protein, % of DM	10.0	10.0
Rumen degradable protein, % of CP	85	73
Rumen undegradable protein, % of DM	1.8	3.7
Rumen undegradable protein, % of CP	15	27
Neutral detergent fibre (NDF), min % of DM	25-33	25-33
Forage NDF, min % of DM	19-25	19-25
Starch max, % of DM (varies)	15-20	15-20
Macrominerals, % of DM		
Ca	0.31	0.38
P	0.21	0.23
Mg	0.13	0.14
K	0.62	0.68
Na	0.16	0.17
Cl	0.12	0.14
S	0.20	0.20
Trace minerals, mg/kg of DM		
Cu	18	19
Co	0.20	0.20
I	0.58	0.62
Fe	13	15
Mn	38	43
Se	0.3	0.3
Zn	30	32
Vitamins, IU/kg of DM		
Vitamin A	5,850	6,510
Vitamin D	1,595	1,774
Vitamin E	85	177

Table B12: Predicted nutrient concentrations in dietary dry matter needed to meet the requirements for dry, pregnant mature Jersey cows. Energy and protein requirements have been adjusted for changes in energy reserves (-0.24 kg/d) for dry cows at less than 21 days prepartum.

Appendix C: The NASEM Dairy 8 Nutrition Model

The NASEM Dairy 8 publication provides the scientific basis for the NASEM Dairy 8 Nutrition Model, a computer program built around mathematical equations defining nutrient requirements for dairy cattle at all stages of their productive lives. The software along with supporting documentation are free to download at:

 https://nap.nationalacademies.org/catalog/25806/nutrient-requirements-of-dairy-cattle-eighth-revised-edition

Although the NASEM Dairy 8 model can be used to formulate rations, its real purpose is to provide insight into the flow of nutrients from feed into productive processes. The NASEM Dairy 8 publication states: "the model is for evaluation and should be used with caution for formulation". Chapter 20 in the NASEM Dairy 8 publication provides a detailed decription of model equations and the software's help system describes how to use it. Shown below are facsimiles of the main screens, intended only to familiarize the reader with the model's structure and user interface.

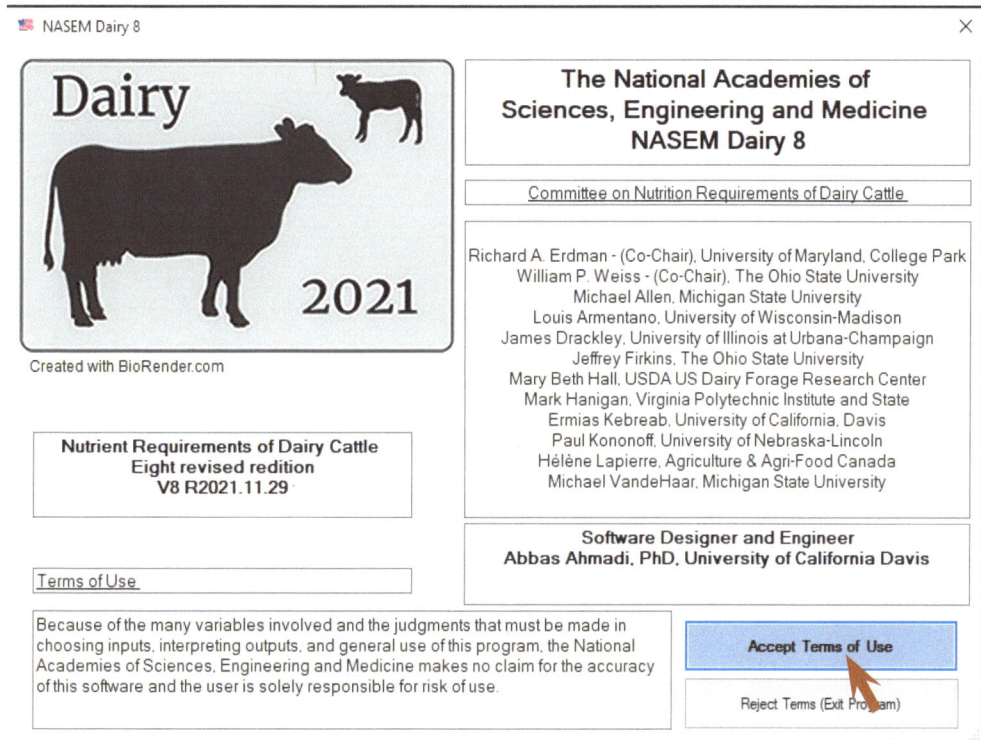

Screen 1

NASEM Dairy 8: Nutrient Requirements of Dairy Cattle V8 R2022.02.15 [SM Lactation 150]

File   Go To...   Help

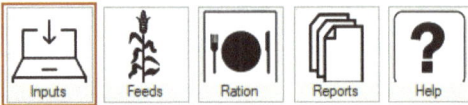

[Inputs] [Feeds] [Ration] [Reports] [Help]

(Program Settings)   Animal Description/Management   Production

Units

- ⦿ Metric
- ○ British Imperial

Basis

- ⦿ Dry Matter
- ○ As Fed

Comments

You can enter a short comment about the current simulation in this field.

Reset Simulation Directory to Factory Default

Clean up Output and Xls folders

Remember ration results for this animal type

Load remembered ration results for this animal type

Ration Results Sidebar

MP Supplied

MP Sup|
Diet NE|
Absorb|
Predict|
NEL Ba|
NP bal|
Diet MP|
Diet RD|
Diet RU|
Absorb|
Diet NE|
Diet Fo|
Diet St|
☐ Use De

Dropdown list:
- Diet Starch
- Diet Unsaturated Fatty Acids
- Energy Allowable ADG w/o Conceptus
- Energy Allowable ADG with Conceptus
- Energy Efficiency (NE/GE)
- Entered RHA Milk Protein
- Manure excretion
- ME Balance
- Methane production
- MiN Outflow
- MP Allowable milk
- **MP Balance**
- MP Supplied
- MP Supplied/ME Supplied
- NEL allowable milk
- NEL Balance
- Nitrogen excretion
- NP balance
- Phosphorus excretion
- Pred Milk Prot: % of maximum milk protein
- Predicted Milk Fat
- Predicted Milk Production
- Predicted Milk Protein
- Predicted water Intake
- Protein allowable growth
- RDP Balance
- Target ADG
- Target ADG with Conceptus
- Target Milk Production
- Volatile Solids excretion

Screen 2

NASEM Dairy 8: Nutrient Requirements of Dairy Cattle V8 R2022.02.15 [SM Lactation 200]

File   Go To...   Help

[Inputs] [Feeds] [Ration] [Reports] [Help]

Program Settings   (Animal Description/Management)   Production

Animal Type	Lactating Cow	▾	Percent First Parity	33	(0-100)
Animal Breed	Holstein	▾	Days in Milk	200	days
Mature Weight	700	kg	Age At First Calving	24.000	months
☑ Compute Mature Weight from the Breed			Days Pregnant	100	days
Age	56.000	months	Temperature	0.00	deg C
Body Weight	700	kg			
Condition Score	3.0	(1-5)			

Grazing
- ○ Grazing
- ⦿ No Grazing

Topography   Mild

Distance Between Pasture and Milking   N/A   km

One-Way Trips   N/A   /d

Temperature only applies to young calves. For dry and lactating cows, environmental temperature is only used to estimate water intake and nothing else.

Warning - If cows are grazing, you must also include a feed from Pasture category in ration to obtain estimated grazing requirements. If cows are not grazing ensure that no feeds from Pasture category are in the diet.

114

NASEM Dairy 8: Nutrient Requirements of Dairy Cattle V8 R2022.02.15 [SM Lactation 200]

File   Go To...   Help

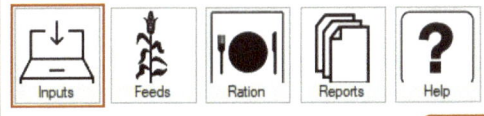

[Inputs] [Feeds] [Ration] [Reports] [Help]

Program Settings   Animal Description/Management   (Production)

Calf Birth Weight: 44.100   kg/d

☑ Compute Calf Birth Weight from Mature Weight and Parity

Growth Rate   0.10   kg/d

Body Reserve   0.00   None
Replenishment
Rate

Milk Production   42.000   kg

Times Milked per Day   3.0   times

☑ Compute Milk Components from the Breed

Milk Components
    Milk Fat   3.80   %

    Milk Protein
        ○ Crude Protein
        ● True Protein   3.10   %

    Milk Lactose   4.85   %

Milk Protein: Rolling Herd Average (RHA)
        ○ Crude Protein
        ● True Protein   396   kg/305 d

NASEM Dairy 8: Nutrient Requirements of Dairy Cattle V8 R2022.02.15 [SM Lactation 200]

File   Go To...   Help

[Inputs] [Feeds] [Ration] [Reports] [Help]

Feeds

Feeds

Barley grain, steam rolled
Barley silage, mid-maturity
Beet pulp, dry
Blood meal, high dRUP
Canola meal
Corn grain, steam-flaked
DDGS, high protein
Calcium soaps
Grass lg mixt, leg., hay, immtr

[∧]
[∨]

Add Feeds to Ration

Remove Feed From Ration

Clear Ration

Save Feed in Feed Library

Remove Feed from Feed Library

Feed Components and Nutrients

Component	Value
Feed Library	NRC 2020
UID	NRC16F1074
Index	258
Name	Barley grain, steam rolled
Category	Energy Source
Type	Concentrate
DM (% As Fed)	88.742
Concentration (%)	100
Locked	0
DE, Base (Mcal/kg)	3.435
ADF (% DM)	7.332
NDF (% DM)	18.634
48 h in Vitro NDF Digestibility (% ...	51.545
Lignin (% DM)	1.724

Edit Feed Components and Nutrients

Use in vitro NDF digest to estimate energy: Do not use   ▾

Feeding Monensin at 250 to 450 mg/day (Yes/No)?   No   ▾

Warning: For your changes to take effect, you must leave the Feeds screen and go to the Ration screen.

Select one or more feeds from NRC dairy feed library

Select a feed category

All

Select one or more feeds

- Corn germ
- Corn germ meal
- Corn gluten feed, dry
- Corn gluten feed, wet
- Corn gluten meal
- Corn grain dry, coarse grind
- Corn grain dry, fine grind
- Corn grain dry, medium grind
- Corn grain HM, coarse grind
- Corn grain HM, fine grind
- Corn grain screenings
- **Corn grain, steam-flaked**
- Corn hominy feed
- Corn silage, immature

Selected ingredients and their profile:

	Feed Library	UID	Index	Name	Category	Type	DM (% As Fed)	Concentration (%)
	NRC 2020	NRC16F12	12	Barley silage, mid...	Grain Crop Forage	Forage	36.12	0
	NRC 2020	NRC16F14	14	Beet pulp, dry	By-Product/Other	Concentrate	92.298	100
	NRC 2020	NRC16F1000	176	Blood meal, high ...	Animal Protein	Concentrate	90.864	100
	NRC 2020	NRC16F25	25	Calcium soaps	Fatty Acid Supple...	Concentrate	95.3	100
	NRC 2020	NRC16F28	28	Canola meal	Plant Protein	Concentrate	89.128	100
►	NRC 2020	NRC16F46	46	Corn grain, steam...	Energy Source	Concentrate	85.74	100

To delete a feed in the selected ingredient list, select that row and press the DEL key on your keyboard.

Cancel   Add selected ingredients to the ration

NASEM Dairy 8: Nutrient Requirements of Dairy Cattle V8 R2022.02.15 [SM Lactation 200]

File   Go To...   Help

Inputs   Feeds   Ration   Reports   Help

Ration

Ration List (Dry Matter Basis)

Feed prices are in $/Metric ton as fed, but the Diet Cost is in $/day

NO	Fd Name	Qty (kg/day)	% Total	Price
1	Barley grain, steam rolled	8.000000	30.769231	0.0
2	Barley silage, mid-maturity	7.500000	28.846154	0.0
3	Beet pulp, dry	1.000000	3.846154	0.0
4	Blood meal, high dRUP	0.500000	1.923077	0.0
5	Canola meal	2.500000	9.615385	0.0
6	Corn grain, steam-flaked	2.000000	7.692308	0.0
7	DDGS, high protein	2.000000	7.692308	0.0
8	Calcium soaps	0.500000	1.923077	0.0
9	Grass lg mixt, leg., hay, immtr	2.000000	7.692308	0.0
	Totals	26.000	100.000	0.000

Ration Results	Value	Unit
MP Balance	136.86	g/d
MP Supplied/ME Supplied	38.98	g/Mcal
Diet NEL	1.78	Mcal/kg
Absorbed Arg	136.45	g/d
Predicted Milk Protein	1219.76	g/d
NEL Balance	6.83	Mcal/d
NP balance	-1.52	g/d
Diet MP	10.51	% of DM
Diet RDP	12.01	% of DM
Diet RUP	6.28	% of DM
Absorbed Met	59.51	g/d
Diet NDF	32.51	% of DM
Diet ForNDF	18.63	% DM

Total Intake

26.000   kg/day

Estimated Intake Based on Animal

26.392   kg/day   Use this Estimate

Estimated Intake Based on Animal/Fiber

25.737   kg/day   Use this Estimate

Set to 100%

Refresh Sidebar

Warning

When the [Total Intake] slot is zero, do not enter any value for any feed in the grid. You must first enter a non-zero value in this slot and then start filling the grid.

NASEM Dairy 8: Nutrient Requirements of Dairy Cattle V8 R2022.02.15 [SM Lactation 200]

File   Go To...   Help

[Inputs] [Feeds] [Ration] [Reports] [Help]

Reports

Select one or more reports

- ☐ 01. Animal Inputs
- ☑ 02. Diet Summary
- ☐ 03. Ingredient Macro-Nutrient Contributions
- ☐ 04. Energy Supply
- ☐ 05. Fatty Acid Supply
- ☐ 06. Protein and Amino Acid Supply and Requirements
- ☐ 07. Mineral and Vitamin Supply and Requirements
- ☐ 08. Environmental Impact
- ☐ 09. Ingredient Mineral Contributions
- ☐ 10. All

Generate Selected Reports

Generate Excel Spreadsheet for Complete Feed Composition of Ingredients in Diet

---

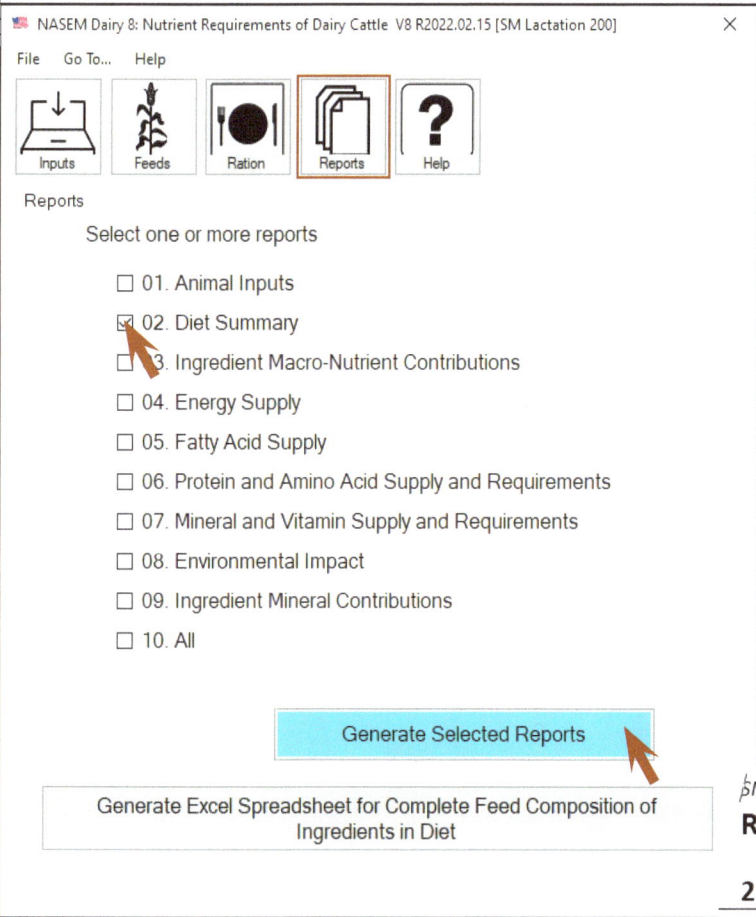

SM Lactation 200

Report 2. Diet Summary (DM Basis)

2.1 Macro-nutrients

Nutrient	Content
Dry Matter, %	62.6
Forage, % DM	36.5
CP, % DM	18.3
ME, Mcal/kg	2.70
MP, % DM	10.51
NEL, Mcal/kg	1.78
RUP, Base, % DM	6.3
RDP, % DM	12.0
Dig. RUP, % DM	5.0
ADF, % DM	19.3
NDF, % DM	32.5
ADF/NDF, Ratio	0.59
Forage NDF, % DM	18.6
Starch, % DM	27.3
WSC, % DM	5.7
Ash, % DM	5.6
Total FA, % DM	3.69
Ca, % DM	0.39
P, % DM	0.42
Mg, % DM	0.22
K, % DM	1.18
Na, % DM	0.10
Cl, % DM	0.32
S, % DM	0.26
DCAD, mEq/kg	90
Cost, $/ton As Fed	0.00
Cost, $/day	0.00

Index

D

www.ingramcontent.com/pod-product-compliance
Lightning Source LLC
Chambersburg PA
CBHW042058210326
41597CB00045B/61